AF540416

# Biotechnology:

## *Theory and Techniques*

Parmeshwar Hegde

MAXFORD BOOKS
New Delhi-(INDIA)

Reprint : 2014



**Biotechnology: Theory and Techniques**

ISBN: 81-8116-058-4

First Edition 2008

*Published by :*

**MAXFORD BOOKS**

**Sales office:**

4264/3, IInd Floor, Ansari Road, Darya Ganj,

New Delhi–110 002, Phone.: 011-65156284

**Registered office:**

95 Medha Appartments Mayur Vihar Phase I,

Delhi–110 091 Phone: 011-22743537,

Typesetting : Kingston

# Contents...

# *Preface...*

Biotechnology is technique which involves the application of biological organism or their components, systems or processes in manufacturing and service industries to make or modify products, to improve plants or animals or to develop micro-organism for any special purpose.

Since 1953, when James Watson and Francis Crick identified the structure of deoxyribonucleic acid (DNA) as the genetic basis of all living organism, the scientific understanding of biological and genetic processes has dramatically accelerated.

Biotechnology revolution has spawned new industries focused on manipulating human, animal, plant and microbial agents to create heretofore unattainable products and services. Biotechnology occupies a very strategic positions in the socio-economic advancement and development of the nation in particular and world at large.

The book, where on the one hand is to serve the purpose as the source of the updated informations in the concerned academic field, there on the other for a general reader it will disclose the secrets of the sophistication in the products and services at his disposal today.

**Author**

CHAPTER-1

# Introduction

The word 'biotechnology' was coined by Karl Ereky, a Hungarian engineer, in 1919 to refer to methods and techniques that allow the production of substances from raw materials with the aid of living organisms. A standard definition of biotechnology has been reached in the Convention on Biological Diversity (1992): "Any technological application that uses biological systems, living organisms or derivatives thereof, to make or modify products and processes for specific use". This definition was agreed by 168 member nations, and also accepted by the Food and Agricultural Organization of the United Nations and the World Health Organization (WHO).

Biotechnologies are therefore a collection of techniques or processes using living organisms or their units to develop added-value products and services. When applied at industrial and commercial scale, biotechnologies give rise to bio-industries. Conventional biotechnologies include plant and animal breeding, the use of microorganisms and enzymes in fermentations, preparation and preservation of products, as well in the control pests (e.g. integrated pest control). More advanced biotechnologies mainly refer to the use of recombinant deoxyribonucleic acid (DNA) techniques (i.e. the identification, splicing and transfer of genes from one organism to another), which are now supported by the research on genetic information (genomics). This distinction is just a convenient one, as modern techniques are used to empower conventional methods, e.g. recombinant enzymes and genetic markers have been employed to improve fermentations, plant and animal breeding. It is, however, true that the wide range of biotechnologies, from the most simple to the very sophisticated ones, allows each country to select those biotechnologies which suit its needs, development priorities, and by doing so could even reach a level of excellence (e.g. the case of developing countries who have used *in vitro* micropropagation and plant-tissue cultures to become world leading exporters of flowers and commodities).

The potential of biotechnology to contribute to increasing agricultural, food and feed production, improving human and animal health, and abating pollution and protecting the environment, has been acknowledged in *Agenda 21* – the work programme adopted by the 1992 United Nations Conference on Environment and Development in Rio de Janeiro. In 2001, the Human Development Report considered biotechnology as the means to tackle major health challenges in the poor countries, such as infectious diseases (tuberculosis), malaria and HIV/AIDS, and as an adequate tool to deal with the development of the regions left behind by the 'green revolution', but home to more than half of the world's poorest populations, depending on agriculture, agroforestry and livestock husbandry. New and more effective vaccines, drugs and diagnostic tools, as well as more food and feed of high nutritional value will be needed to meet the expanding needs of the world's populations.

Biotechnology and bio-industry are becoming an integral part of the knowledge-based economy, because they are closely associated with the progress in life sciences, and the applied

sciences and technologies linked to them. A new model of economic activity is being ushered – bio-economy – whereby new types of enterprises are created and old industries are revitalized. Bio-economy is defined as including all industries, economic activities and interests organized around living systems. Bio-economy can be divided into two primary industry segments: the bio-resource industries that can directly exploit biotic resources – crop production, horticulture, forestry, livestock and poultry, aquaculture and fisheries –, and the related industries that have large stakes as either suppliers or customers to the bio-resource sector – agro-chemicals and seeds, biotechnologies and bio-industry, energy, food and fiber processing and retailing, pharmaceuticals and health care, banking and insurance. All these industries are closely associated with the economic impact of human-induced change to biological systems.

The potential of this bio-economy to spur economic growth and create wealth, through enhancing industrial productivity, is unprecedented. It is therefore no surprise that high income and technologically-advanced countries have made huge investments in research and development (R&D) in the life sciences, biotechnology and bio-industry.

In 2001, bio-industry was estimated to have generated $34.8 billion in revenues and employed about 190,000 persons in publicly-traded firms, worldwide. These are impressive results given that, in 1992, bio-industry was estimated to have generated $8.1 billion and employed less than 100,000 persons.

The main beneficiaries of the current 'biotechnology revolution' and derived bio-industries are largely the industrialized and technologically-advanced countries, i.e. those which enjoy a large investment of their domestic product in R&D

and technological innovation. Thus, the USA, Canada and Europe account for about 97% of the global biotechnology revenues, 96% of persons employed in biotechnology ventures and 88% of the total biotechnology firms. Ensuring that those who need biotechnology have access to it remains 'therefore' a major challenge. Similarly, creating a conducive environment for the acquisition, adaptation and diffusion of biotechnology in developing countries is another great challenge. However, a number of developing countries are increasingly using biotechnology and have created a successful bio-industry, at the same time as they are raising their investments in R&D in the life sciences.

## US biotechnology and bio-industry

According to Frost & Sullivan Chemicals Group, Oxford, United Kingdom, nearly 4,400 biotechnology companies were active globally in 2003: 1,850 (43%) in North America; 1,875 (43%) in Europe; 380 (9%) in Asia; and 200 (5%) in Australia. These companies cover the gamut from pure R&D participants to integrated manufacturers to contract manufacturing organizations (CMOs) The USA leads with the largest number of registered biotechnology companies in the world (318) , followed by Europe (102). Annual turnover (2002) of these companies has been $33 billion in the USA and only $12.8 billion in Europe. Some $20.5 billion had been allocated to research in the USA, compared with $7.6 billion in Europe.

Ernst & Young – a consultancy firm – makes a difference between US companies which have medicines and the others. The former include pioneers such as Amgen, Inc., Genentech, Inc., Genzyme Corporation, Chiron Corp., Biogen, Inc. These five companies have an annual turnover representing one-third of

the sector's total, i.e. $11.6 billion out of $33 billion; in addition, their product portfolio enables them, with respect to their turnover and stock value, to compete with the big pharmaceutical groups. For instance, Amgen, Inc., with a $75-billion market capitalization, is more important than Eli Lilly & Co., while Genentech, Inc.'s market capitalization is twice as big as that of Bayer AG.

In 2002, Amgen, Inc., had six products on the market with global revenues amounting to $4,991 million. With 11 products on the market and revenues worth $2,164 million, Genentech, Inc., followed in second place. The remaining places in the top five were filled out by Serono SA (six products, $1,423 million), Biogen, Inc. (two products, $1,034 million) and Genzyme Corporation (five products, $858 million).

Over the last decade, a clutch of companies has amassed significant profits from a relatively limited portfolio of drugs. There is, today, heightened recognition that lucrative opportunities await companies that can develop even a single live-saving biotechnology drug. For instance, Amgen, Inc.'s revenues increased by over 40% from 2001 to 2002 on the $2 billion Amgen made in 2002 from sales of Epogen and the $1.5 billion earned from the sales of Neupogen. Over $1 billion in sales of Rituxan – monoclonal antibody against cancer – in 2002 helped Genentech, Inc., record a 25% growth over its 2001 performance.

In California, there are two biotechnology 'clusters' of global importance: one at San Diego-La Jolla, south of Los Angeles, and the other at Bay Area, near San Francisco. A cluster is defined as an interconnection of enterprises and institutions in a precise sector of knowledge, geographically close to each other and networked through all kinds of links, starting with those concerning clients and suppliers. In both biotechnology clusters, it does not take more than 10 minutes to move from one company to the other. Biocom – a powerful ensemble of 450 enterprises, including about 400 in biotechnology, in the region of San Diego – is helping the San Diego cluster in all aspects of its functioning, which also includes lobbying the politicians and the various actors of the bio-economy. The cluster relies on the intensity of exchanges between industry managers and university research centres; for instance, one of the objectives is to shorten the average time needed to set up a licensing contract between a university and a biotechnology company: it generally takes 10 months to establish such a contract, which is considered too long, and the cluster association gathers all the stakeholders to discuss the relevant matters and conclude rapidly.

The clusters have developed the proof of concept, i.e. to try to show that behind an idea, a theory or a concept, there could be a business model and eventually a blockbuster drug. Such an endeavour made by the researchers toward the industry would lead to licensing agreements which would reward the discovery work. The strategic alliance between politics, basic research and pharmaceutical industry (including biotechnology or not) within the cluster would be meaningless without capital. In fact, the success of bio-industry is above all associated with an efficient capital market, according to David Pyott, chief executive officer of Allergan, the world leader of ophtalmic products and unique owner of Botox – a product used in esthetic surgery and the main source of the company's wealth. There cannot be any cluster without a dense network of investors, business angels, venture capitalists and bankers, ready to take part in companies being constituted (Mamou, 2004*e*).

Besides the two Californian clusters which represented 25.6% of US companies in 2001, the following percentages correspond to the other States: 8.6% – Massachusetts; 7.7% – Maryland; 5.9% – New Jersey; 5.8% – North Carolina; 4.6% – Pennsylvania; 3.4% – Texas; 3.1% – Washington; 3.1% – New York; 2.5% – Wisconsin; and 29.7% for the rest of the country (data from the US Department of Commerce Technology Administration and Bureau of Industry and Security).

## Europe's biotechnology and bio-industry

The European bio-industry is less mature than its 25-year-old US counterpart. Actelion of Switzerland qualified as the world's fastest-growing drugs group in sales terms following the launch of its first drug, Tracleer, but it had just only achieved profitability in 2003. Similarly, barely a few European biotechnology companies earn money. Only Serono SA – the Swiss powerhouse of European biotechnology that grew out of a hormone extraction business with a 50-year record of profitability – has a market capitalization to rival US leaders. Serono SA, which is the world leader in the treatment of infertility and also well known in endocrinology and the treatment of multiple sclerosis, had made in 2002 a $333 million net profit from $1,546 million of sales, and devoted 23% of these sales to its research-and-development division where 1,200 people were working. The Spanish subsidiary of Serono SA in Madrid is now producing recombinant human growth hormone for the whole world, while the factories in the USA and Switzerland ceased to produce it. The Spanish subsidiary had to invest €36 million in order to raise its production, as well as another 5 million to upgrade its installations to the production of other recombinant pharmaceuticals to be exported worldwide.

In spite of a wealth of world-class science, the picture in

much of Europe is of an industry that lacks the scale to compete and faces financial crunch, which may force many too seek mergers with stronger rivals.

Germany has overtaken the United Kingdom and France, and is currently home to more biotechnology companies than any country except the USA. But far from pushing the boundaries of biomedical sciences, many companies are putting cutting-edge research on hold and selling valuable technology just to stay solvent. Until the mid- 1990s, legislation on genetic engineering in effect ruled out the building of a German bio-industry. Since then, the more than 400 companies set up in Germany needed to raise at least $496 million from venture capitalists over 2004 to refinance their hunt for new medicines, according to Ernst & Young. Most were far from having profitable products and, with stock markets in effect closed to biotechnology companies following the bursting of the bubble in 2000, they were left to seek fourth or even fifth rounds of private financing.

The biggest German biotechnology companies, such as GPC Biotech and Medigene, were able to raise significant sums in initial public offerings at the peak of Neuer Markt, Germany's market for growth stocks. But when the technology bubble burst in 2000, it became clear to GPC Biotech that investors put very little value on 'blue-sky' research. 'They wanted to see proven drug candidates in clinical trials', states Mirko Scherer, chief financial officer. The only option for companies such as GPC Biotech and Medigene was to buy drugs that could be brought to market more quickly. GPC Biotech has used the cash it earned from setting up a research centre for Altana, the German chemicals and pharmaceutical group, to acquire the rights to Satraplatin, a cancer treatment that was in the late stages of development. In October 2003, regulators gave the authorization to initiate the last round of clinical trials. After a series of clinical setbacks, Medigene has moth-balled its early-stage research to cut costs and licensed-in late-stage products to make up for two of its own drugs that failed. The strategy will help the company eke out its cash; but cutting back on research will leave its pipeline looking thin.

Many of Germany's biotechnology companies have abandoned ambitious plans to develop their own products and chosen instead to license their drug leads to big pharmaceutical companies in exchange for funding that will allow them to continue their research. This approach is supported by the acute shortage of potential new medicines in development by the world's biggest pharmaceutical companies. But Germany's bio-industry has few experimental drugs to sell – about 15 compared with the more than 150 in the United Kingdom's more established industry. Moreover, most of Germany's experimental drugs are in the early stages of development, when the probability of failure is as high as 90%. That reduces the price that pharmaceutical companies are willing to pay for them.

Companies also have to struggle with less flexible corporate rules than their rivals in the United Kingdom and the USA. Listed companies complain that the Frankfurt stock exchange does not allow injections of private equity, common in US biotechnology. As a result, few of Germany's private companies state they expect to float in Frankfurt. Most are looking to the USA, the United Kingdom or Switzerland, where investors are more comfortable with high-risk stocks. But many German companies may not survive long enough to make the choice.

Faced with this bleak outlook, many in the industry agree that the only solution is a wave of consolidation that will have fewer, larger companies with more diverse development pipelines. A number of investors in Germany's bio-industry are already pushing in this direction. TVM, the leading German venture capital group, had stakes in 14 German biotechnology companies and was trying to merge most of them. TVM sold off all Cardion's drug leads after failing to find a merger partner for the arthritis and transplant medicine specialists. After raising $14.1 million in 2002, Cardion has become a shell company that may one day earn royalties if its discoveries make it to market. United Kingdom-based Apax Partners was said to have put almost its entire German portfolio up for sale. The fate or MetaGene Pharmaceuticals, one of Apax's companies, may show what awaits many others. In October 2003, the company was bought by the British Astex, which planned to close the German operations after stripping out their best science and $15-million bank balance.

GPS Biotech's chief financial officer was critical of the investors who turned their backs on Germany and put 90% of their funds in the USA, while a lot of European companies were very cheap. And although Stephan Weselau, chief financial officer of Xantos, was frustrated that venture capitalists see little value in his young company's anti-cancer technology, he was adamant about the need for Germany's emerging biotechnology to consolidate if it was to compete against established companies in Boston and San Diego.

In the United Kingdom, the market for initial public offerings has been all but closed to biotechnology for the three-year period 2000-2002, while it has reopened in the USA in 2003. City of London institutions, many of them made huge losses on biotechnology, were reluctant to back new issues and have become more fussy about which quoted companies they were prepared to finance. The United Kingdom is home to a third of Europe's 1,500 biotechnology companies and more than 40% of its products in development. But although the United Kingdom had 38 marketed biotechnology products and seven more medicines awaiting approval by the end of 2003, analysts stated they were too few genuine blockbusters with the sort of sales potential needed to attract investors' attention away from the USA. A dramatic case is that of PPL (Pharmaceutical Proteins Ltd) Therapeutics – the company set up to produce drugs in the milk of genetically-engineered sheep. By mid-December 2003, the company raised a paltry \$295,000 when auctioneers put a mixed catalogue of redundant farm machinery and laboratory equipment under the hammer. This case proved that exciting research (Dolly and Polly sheep) did not always lead to commercial success.

The profitable British companies reported pre-tax profits of £145 million in 2003, less than 15% of the \$1.9-billion pre-tax profits reported by the US Amgen. By mid-2003, the British biotechnology sector seemed to be coming of age. Investors could choose between three companies that had successfully launched several products and boasted market capitalizations in excess of \$884 million. Since then they saw PowderJect Pharmaceuticals plc be acquired by Chiron Corp., the US vaccines group, in May 2003 for a deal value of £542 million; General Electric swoop in with a £5.7 billion bid for Amersham, the diagnostics and biotechnology company, in October 2003. Earlier on, in July 2000, Oxford Asymmetry was purchased by the German company Evotec Biosystems for £343 million, and in September 2002, Rosemont Pharma has been acquired by the US firm Bio-Technology General for £64 million.

In May 2004, Union Chimique Belge (UCB) has agreed to buy Celltech, the United Kingdom's biggest biotechnology company, for £1.53 billion or €2.26 billion. The surprise acquisition was accompanied by a licensing deal that gives UCB the rights to Celltech's new treatment for rheumatoid arthritis. This drug – CPD 870 – with forecast annual sales of more than $1 billion would account for about half the company's valuation. Goran Ando, the Celltech chief executive, who will become deputy chief executive of UCB, stated: "We will immediately have the financial wherewithal, the global commercial reach and the R&D strength to take all our drugs to market". News of the deal, which will be funded with debt, sent Celltech shares 26% higher to £5.42, while UCB shares fell 4% to €33.68.

Celltech has been the grandfather of the British biotechnology sector since it has been founded in 1980. With a mixture of seed funding from the Thatcher government and the private sector, the company was set up to commercialize the discovery of monoclonal antibodies that can become powerful medicines. Listed in 1993, the company made steady progress in its own research operations, but only gained products and financial stability with the acquisitions of Chiroscience in 1999 and Medeva in 2000. It also acquired Oxford GlycoSciences in May 2003 for a deal value of £140 million. The great hopes Celltech has generated were based largely on CDP 870, the arthritis drug it planned to bring to market in 2007 and which could be by far the best-selling product to come out of a British biotechnology company. After the UCB-Celltech deal, the group ranked fifth among the top five biopharmaceutical companies, i.e. behind: Amgen, Inc., €6.6 billion of revenue in 2003; Novo Nordisk, €3.6 billion; Schering, €3.5 billion; and Genentech, Inc., €2.6 billion.

Celltech is the biggest acquisition by the Belgian group UCB, which branched out from heavy chemicals only in the 1980s. Its chief executive since 1987, Georges Jacob, stated that when he joined UCB he found a company 'devoted to chemicals, dominated by engineers, pretty old-fashioned and very much part of heavy industry'. Built entirely on internal growth, UCB's only other sizeable acquisition was its purchase of the speciality chemicals business of USA-based Solutia in December 2002 for $500 million, a move that split the Belgian group's €3 billion revenues evenly between pharmaceuticals and chemicals. One constant was the continued presence of a powerful family shareholder, owning 40% of UCB's equity via a complicated holding structure (Firn and Minder, UCB made its first foray into pharmaceuticals in the 1950s with the development of a molecule it sold to Pfizer, Inc., and became Atarax, an anti-histamine used to relieve anxiety. The relationship with Pfizer, Inc., was revived in a more lucrative fashion for UCB following the 1987 launch of Zyrtec, a blockbuster allergy treatment Pfizer, Inc., helped distribute in the USA. Although UCB has a follow-up drug to Zyrtec, it faces the loss of the US patent in 2007. UCB had also to fight patent challenges to its other main drug, Keppra, an epilepsy treatment. With the takeover of Celltech, the group that will emerge will be one of Europe's leading biotechnology companies; the deal will give UCB a pipeline of antibody treatments for cancer and inflammatory diseases to add to its allergy and epilepsy medicines. The expansion in health-care activities would lead the group to divest its remaining chemical business, according to most analysts.

Based on 2003 results, the combined market capitalization of UCB Pharma and Celltech will reach €7.14 billion; revenues, €2,121 million; earnings before interest, tax and amortization, €472 million; pharmaceutical research-and-development budget, €397 million; the number of employees is about 1,450. UCB decided Celltech could be its stepping stone into biotechnology after entering an auction for the marketing rights to Celltech's

CDP 870 against arthritis touted as a $1 billion-a-year blockbuster drug. After seeing trial data not revealed to the wider market, it decided to buy the whole company. In the United Kingdom's biotechnology sector, after this takeover and following the earlier acquisition of PowderJect Pharmaceuticals plc and Amersham by US companies, there is not much left except Acambis, another vaccine-maker, valued at about £325 million, and a string of companies below the £200 million mark where liquidity can be a problem for investors. The industry was 'therefore' afraid it would be swamped by its much larger rivals.

Martyn Postle, director of Cambridge Healthcare and Biotech, a consultancy, stated: "We could end up with the UK performing the role of the research division of US multinationals". According to the head of the Bioindustry Association (BIA), "It is clearly the fact that US companies are able to raise much, much more money than in the United Kingdom, which puts them in a much stronger position". The BIA called for changes in the rules on 'pre-emption rights', which give existing shareholders priority in secondary equity offerings. As Celltech was by far the most liquid stock in the sector, there could be a broader impact on the way the financial sector treat biotechnology, including a reduction in the number of specialist investors and analysts covering the sector.

It is important for the United Kingdom to create an environment in which biotechnology can flourish. The industry has called for institutional reform, including measures to make it easier for companies to raise new capital. The British government must also ensure that its higher education system continues to produce world-class scientists. That reinforces the need for reforms to boost funding of universities. The Celltech takeover need not to be seen as a national defeat for the United Kingdom. The combined company may end up being listed in London. Even if it does not, Celltech's research base in the United Kingdom will expand. Its investors have been rewarded for their faith and, if its CDP 870 drug is approved, UCB's shareholders will also benefit. But for Celltech's executives, the acquisition is a victory for Europe. The takeover creates an innovative European biotechnology company that is big enough, and has sufficient financial resources, to compete globally. 'The key was to have viable European businesses that have a sustainable long-term presence', stated Goran Ando, who confirmed that UCB's research will be run from Celltech's old base in Slough. A lot of hopes ride on the success of UCB and Celltech, that will allow the fledgling bio-industry to thrive in Europe and prevent the life sciences to migrate to the USA.

In France, according to the association France Biotech, there were in 2003270 biotechnology companies, i.e. focused on the life sciences and being less than 25-years old, that employed 4,500 people – a number that could be multiplied by four or five if about €3 billion were to be invested in public research over three years. France was investing only €300 million of private funds and €100 million of public funds in biotechnology, far behind Germany and the United Kingdom which invested about €900 million per year, each. In 2003, France launched a five-year Biotech Plan that aimed at restoring the visibility and attractivity of France in 2008-2010. Three areas – human health, agrifood and environment – were expected to attract the funds as well as the efforts of universities, public and private laboratories, hospitals, enterprises and investors.

SangStat, a biotechnology company created in 1989 in the Silicon Valley by Philippe Pouletty – a French medical immunologist – is working on organ transplants. It was established in San Francisco and Lyon. At the time of its creation, venture capital was just starting in France to support such endeavour in biotechnologies. Between 600 million and 2 billion Francs were needed to set up a biotechnology corporation, which will develop one and perhaps two new drugs. Thus, instead of creating the company in France, were bankruptcy was very probable, the company was set up in California. A second corporation, (DAS) was established in 1994 by the same French medical scientist, when venture capital in Europe became a current practice, so that two companies were created at the same time: DAS France and DAS US in San Francisco, both belonging to the same group and having the same shareholders. Being established in Europe and the USA, a higher flexibility could be achieved from the financial viewpoint and a better resilience versus stock exchange fluctuations. Drug Abuse Sciences (DAS) has been able, in 1999, to increase its capital by 140 million Francs (€21.3 million) with the help of European investors. DAS was specialized in drug abuse and alcoholism. Its original approach was to study the neurological disorders of the patient so as to promote abstinence, treat overdoses and prevent dependence through new therapies. P. Pouletty, in 1994, made a survey of existing biotechnology companies and found that hundreds of them were working on cancer, dozens on gene therapy, diabetes, etc., but not a single one out of 1,300 surveyed companies was working on drug and alcohol addiction. Even the big pharmaceutical groups had no significant activity in this area, while drug and alcohol addiction is considered the greatest problem of public health in industrialized countries.

For instance, in France, 2.5% of the annual gross domestic product is devoted to these illnesses, and some $250 billion in the USA. A first product, Naltrel, improves the current treatment of alcoholism by naltrexone. The latter, to be efficient, must be taken as pills every day. But few alcoholics can strictly follow this kind of treatment. In order to enable the patients to free themselves 15 from this daily constraint, a single monthly intramuscular injection of a delayed-action microencapsulated product has been developed, and helps alcoholics and drug-addicts to abstain from taking their drug. The molecule developed inhibits the receptors in the brain which are stimulated by opium-related substances. Another successful product, named COC-AB, has been developed for the urgent treatment of cocaine overdoses.

This molecule recognizes cocaine in the bloodstream and traps it before it reaches the brain, and then it is excreted through the kidneys in urine. Commercialization of the medicine was expected to help 250,000 cocaine-addicts who are admitted annually in the medical emergency services. In the long term, DAS intends to develop preventive compounds which can inhibit the penetration of the drug into the brain. DAS was expected to become by 2005-2007 a world leading pharmaceutical company in the treatment of alcoholism and drug addiction or abuse. This forecast was based on the current figures: 30 million chronic patients in the USA and Europe, including 22 million of alcoholics, 6 million of cocaine- and 2 million of heroine-addicted persons (Lorelle, 1999*a*).

SangStat is a world leader in the treatment of the rejection of organ transplants and intends to extend its expertise and know-how to the whole area of transplantation; two drugs in the USA and three in Europe were already marketed. Another success story is that of the French biotechnology company, Eurofins, founded in Nantes in 1998 to exploit a patent filed by a couple of researchers from the local faculty of sciences; it currently employs 2,000 people worldwide and increased ten-fold its annual turnover in four years up to €162 million. Its portfolio contains more than 5,000 methods of analyzing biological substances. In

Nantes, where 130 people are working, is located the company's centre of competence that carries out research on the authenticity and origin of foodstuffs. Despite the closing of some sites among the 50 laboratories of Eurofins due to the economy slowdown, and aimed at improving the financial situation, the company wants to remain a growing one. This success story has led Nantes to think of creating a biotechnology city, while giving a strong impetus to medical biotechnology at the town's hospital, where the number of biotechnology researchers soared from 70 to 675. In October 2003, the Institute of Genetics Nantes Atlantique initiated its activities concerning the analysis of human DNA for forensic purposes. This institute which received venture capital from two main sources, will employ 50 people in two-years time in order to meet the demand generated by the extension of the national automated data-base of genetic fingerprinting.

In Spain, Oryzon Genomics is a genomics company based in Madrid, which applies gene discovery techniques to new cereal crops, grapevine and vegetables, as well as to the production of new drugs (especially for Parkinson's and Alzheimer's diseases). It is a young enterprise, an offshoot of the University of Barcelona and the Spanish Council for Scientific Research (CSIC), located in the Scientific Park of Barcelona. With a staff of 22 scientists, the company is experiencing fast growth and is developing an ambitious programme of functional genomics. It has been the first genomics enterprise to have access to special funding from the NEOTEC Programme, in addition to financial support from the Ministry of Science and the Generalitat of Catalonia. Moreover, the National Innovation Enterprise (ENISA), which is part of the General Policy Directorate for the Medium and Small Sized Enterprise of the Ministry of Economy, has invested €400,000 in Oryzon Genomics – this was the first investment made by ENISA in the biotechnology sector. At the end of 2002, Najeti Capital, a venture-capital firm specialized in investments into technology, has acquired 28% of Oryzon Genomics, in order to support the young corporation. The 2003 turnover of the latter was estimated at €500,000, while its clients comprised several agrifood, pharmaceutical companies, and public research centres.

## Japan's biotechnology and bio-industry

Japan is well advanced in plant genetics, and has made breakthroughs in rice genomics. The country is, however, lagging behind the USA on human genetics. Its contribution to the sequencing of the human genome by teams of researchers belonging to the Physics and Chemistry Research Institute of the Agency of Science and Technology as well as to Keio University Medical Department, was about 7%. In order to catch up and to reduce the gap with the USA, the Japanese government has invested important funds in the Millenium Project, launched in April 2000. The project includes three areas: rice genome, human genome and regenerative medicine.

The 2000 budget included 347 billion yens devoted to life sciences. Genomics budget was twice that of neurosciences and amounted to 64 billion yens. Within the framework of the Millenium Project, the Ministry of Health intended to promote the study of genes related to such diseases as cancer, dementia, diabetes and hypertension; results concerning each of these diseases were expected by 2004. The Ministry of International Trade and Industry (MITI) set up a Centre for Analysis of Information Relating to Biological Resources which had a very strong DNA-sequencing capacity, e.g. equivalent to that of Washington University in the USA (sequencing of over 30 million nucleotide pairs per annum), and which will analyze the genome of micro-organisms used in fermentations and provide the information to the industrial sector. In addition, following the project launched in 1999 by Hitachi Ltd, Takeda Chemical Industries and Jutendo Medical Faculty, and aimed at identifying the genetic polymorphisms associated with allergic diseases, a similar project devoted to single nucleotide polymorphisms (SNPs) had been initiated in April 2000 under the aegis of Tokyo University and the Japanese Foundation for Science. The research work is being carried out in a DNA-sequencing centre where 16 private companies send researchers, with a view to contributing to the development of medicines tailored for an individual genetic make-up. This wok is similar to that undertaken by a US-European consortium.

On 30 October 2000, the pharmaceutical group Daiichi Pharmaceutical and the giant electronic company Fujitsu announced an alliance in genomics. Daiichi and Celestar Lexico Science – the biotechnology division of Fujitsu – were pooling their research efforts over the five-year period 2000-2005 to study the genes involved in cancer, ageing, infectious diseases and hypertension. Daiichi devoted about $100 million to this kind of research in 2001-2002, and about 60 scientists were involved in this work of functional genomics.

On 31 January 2003, the Japan Bio-industry Association (JBA) announced that, as of December 2002, the number of 'bioventures' in Japan totalled 334 firms. This announcement was based on a survey – the first of this kind – that the JBA conducted in 2002 to have a better understanding of the nation's bio-industry. A 'bioventure' has been defined as a firm that employs, or develops for, biotechnology applications; that complies with the definition of a small or medium-sized business as prescribed by Japanese law; that has been created 20 years ago; that does not deal primarily in sales or imports/exports.

The three regions with the highest concentrations of bioventures were Kanto (191, i.e. 57% of national total), Kinki/Kansai (55, 16%), and Hokkaido (32, 9.6%). One-third of all ventures (112) were located in Tokyo (inside the Kanto region). The most common field of bioventure operations was pharmaceuticals and diagnostic product development (94 bioventures), followed by customized production of DNA, proteins, etc. (78 bioventures), bio-informatics (41 ventures), and reagents and consumables development (38 bioventures).

The average figures for a bioventure were: 20 employees, including 8.6 staff involved in research and development (R&D), 314 million yens of sales and 153 million yens for R&D costs.

The 334 bioventures employed a total 6,757 employees, including 2,871 R&D staff, had sales amounting to 105 billion yens and R&D costs estimated at 51 billion yen.

## Australia's biotechnology and bio-industry

The consultant firm Ernst & Young has ranked Australia's $12-billion biotechnology and bio-industry as the number one in the Asia and Pacific region and sixth worldwide in its 2003 global biotechnology census. Australia accounts for 67% of public biotechnology revenues for the Asia and Pacific region.

The Australian government has provided a boost to bio-industry by providing close to A$1 billion in public biotechnology expenditure in 2002-2003. There were around 370 companies in Australia in 2002 – an increase from 190 in 2001 – whose core business was biotechnology. Human therapeutics made up 43%, agricultural biotechnology 16% and diagnostics companies 15%. Over 40 companies were listed biotechs and a study released by the Australian Graduate School of Management (Vitale and Sparling) reported that an investment of A$1,000 in each of the

24 biotechnology companies listed on the ASX between 1998 and 2002 would be worth more than A$61,000 in 2003 – an impressive 150% return. During the same period, shares in listed Australian biotechs significantly outperformed those of US biotechs, and the overall performance of listed Australian biotechnology companies was higher than that of the Australian stock market as a whole.

Over A$500 million was raised by Australian listed life-science companies in 2003. The ASX healthcare and biotechnology sector had a market capitalization of A$23.4 billion in 2003, up 18% on 2002. There has been a maturing of the Australian biotechnology sector with greater attention paid to sustainable business models, and identification of unique opportunities that are appealing to investors and partners. The industry is supported by its skilled personnel, with Australia considered to have a greater availability of scientists and engineers than the United Kingdom, Singapore and Germany.

Australia is ranked in the top five countries (with a population of 20 million or more) in terms of the availability of research-and-development (R&D) personnel. It outranks major OECD countries (including USA, Japan, Germany and the United Kingdom) for public expenditure on R&D as a percentage of GDP (Australian Bureau of Statistics 2003). For biomedical R&D, Australia is ranked the second most effective country – in front of the USA, the United Kingdom and Germany – particularly with respect to labour, salaries, utilities and income tax. Australia is ranked third for the cost competitiveness of conducting clinical trials after the Netherlands and Canada. Australian researchers indeed have a strong record of discovering and developing therapeutics. Recent Australian world firsts include the discovery that *Helicobacter pylori* causes gastric ulcers, and the purification and cloning of three of the major regulators of blood-cell transformation – granulocyte colony stimulating factor (GCSF), granulocyte macrophage colony stimulating factor (GMCSF) and leukaemia inhibiting factor (LIF).

Australia is cementing its place at the forefront of stem-cell research with a transparent regulatory system and the establishment of the visionary National Stem Cell Centre (NSCC). An initiative of the Australian government, this centre draws together expertise and infrastructure, and in 2003 it entered into a licensing agreement with the US company LifeCell.

Strong opportunities exist in areas like immunology, reproductive medicine, neurosciences, infectious diseases and cancer. There are also opportunities for bioprospecting given that Australia is home to almost 10% of global plant diversity, with around 80% of plants and microbes in Australia found nowhere else in the world. While 25% of modern medicines come from natural products, it is estimated that only 1% of plants in Australia have been screened for natural compounds. Being the most resilient economy in the world, having the lowest risk of political instability in the world and possessing the most multicultural and multilingual workforce in Asia and the Pacific, Australia had 21 cross-border alliances in 2002, more than France and Switzerland, and 18 more than its nearest Asia-Pacific competitor, according to Ernst & Young's 2003 Beyond Borders report. Australia's geographic location has not been a deterrent for establishing partnerships. All major pharmaceutical companies have a presence in Australia and pharmaceuticals are the third-highest manufactured export for Australia, generating over US $1.5 billion.

The largest drug-discovery partnership in Australia history in 2003, between Merck & Co., Inc. and Melbourne-based Amrad, was valued at up to US $112 million (plus royalties) for the development of drugs against asthma, other respiratory diseases and cancer. It is 'therefore' no wonder the annual US $9.2 billion pharmaceutical industry in Australia is increasingly viewed by the main global players as a valuable source of innovative R&D and technology.

CHAPTER-2

# Biotechnology: Current Achievements and Prospects

In the late 1970s, when the golden era of biotechnology started, the genes for proteins or polypeptides that worked or could work as drugs were cloned in microbial and/or animal cells, and the proteins were produced in bioreactors. Human insulin, human and bovine growth hormones, epidermic growth factor (EGF), erythropoietin (EPO), interferons, anti-haemophilic factors, anti-thrombotic agents (recombinant streptokinase and tissue-plasminogen activator – TPA), anti-hepatitis A and B vaccines, etc., have been produced in this way and successfully commercialized, as well as monoclonal antibodies that fueled and transformed the diagnosis of pathogens and diseases.

It is often stressed that many currently-used medicines have a relative efficiency. For instance, anti-depressants are not effective among 20% to 50% of the patients, betablockers fail in 15% to 35% of the persons treated, and one out of five or even three persons suffering from migraine cannot find a proper medicine to alleviate his/her pain.

It is 'therefore' expected that a personalized medical care with drugs taking account of the genetic make-up of every individual will improve the situation. Thus, in the first pages of the annual report published by Burrill & Company – a Californian Bank specialized in the funding of biotechnologies – Steven Burrill, its chief executive officer, predicted that 'the era of a personalized medicine will generate a market characterized by a small volume per each drug, but the range of products developed for each therapeutic target will be much wider than presently'. There is 'therefore' a strong belief in the effectiveness of a forthcoming individualized medicine, that could even be regenerative (tissue or even organ replacement) or preventive (e.g. it would be possible to anticipate the occurrence of a cancer, rather than to have to try to cure it).

While acknowledging some breakthroughs (e.g. the drug called Gleevec has shown its efficiency against chronic myeloid leukaemia, and Genentech, Inc.'s Avastin can starve tumours by blocking the development of new blood vessels), analysts underline that the transition toward a new therapeutic era is quite slow. In the USA, the $250 billion invested from the late 1960s to 2003 in biotechnologies had a rather low output: out of the 200 most-sold drugs worldwide, only 15% are derived from research and development in the life sciences. In 1996, out of 53 drugs approved for sale worldwide, 9 were derived from biotechnology; in 2000, the figures were 27 and 6; and in 2003, 21 and 14, according to the data provided by the US Food and Drug Administration. Most biotechnology companies continue to spend money in research that does not lead to marketable products. For instance, Vical, after 16 years of research on gene therapy and spending $100 million, has not found a marketable drug. Therefore, the deciphering of the sequence of a gene and of the whole genome of an organism sounds like an attractive short cut, and genomics caught the attention of both the public and the stock markets during the last years of the 20th century. Many new genes have been discovered, with each implying the existence of at least one new protein that might have some therapeutic value.

For instance, an international scientific consortium comprising 58 institutions and laboratories announced the sequencing of almost the whole genome of the rat (*Rattus norvegicus*) in the *Nature* issue dated 1 April 2004. Rats which come from Central Asia, have been widely used as laboratory animals in biological, medical and pharmaceutical research for the last century and half. This animal species became the third mammal after the human species and the mouse the genome of which has been deciphered. The rat's genome is made of 2.75 billion of nucleotide pairs and its size is intermediary between that of the humane genome (2.9 billion nucleotide pairs) and that of the mouse (2.6 billion of nucleotide pairs).

The consortium's work shows that 90% of genes in the rat's genome have their equivalents in the human and murine genomes; such similarity is interpreted as the three species having a common ancestor twenty million years ago. Other genes found in the rat's genome are absent in the two other mammalian species: these are genes involved in the production of pheromones, immune system processes and proteolysis; they are also related to detoxification mechanisms. This is an interesting finding because rats are frequently used to study the potential toxicity for humans of pharmaceuticals and chemicals. The international consortium does not intend to pursue the in-depth study of the rat's genome due to economic reasons. Genome sequencing in mammalian species is being carried out on such species as chimpanzee, macacus, dog, bovine cattle and opposum. On 20 April 2004, a team of 152 researchers working in 67 laboratories and scientific institutions from 11 countries (Australia, Brazil, China, France, Germany, South Africa, South Korea, Sweden, Switzerland, United Kingdom and USA), and coordinated by

Takashi Gojobori (National Genetics Institute of Japan) and Sumio Sugano (University of Tokyo), announced they had identified and described in a detailed manner 21,037 human genes. This group of researchers had been created in 2002 under the name of 'Hinvitational' and their work is a follow-up to the sequencing of the human genome and its overall mapping in 2001. They published their results on internet in the free-access journal *PLos Biology*, edited by the Public Library of Science. By so doing they wish to offer their results to the international scientific community.Starting from the data of the human genome sequencing, they identified the initial and final sequences of each of the 21,037 genes out of the 30,000 to 40,000 that constitute the human genome. Their objective is to draw as many informations as possible on the nature of these genes, their location and their functions, as well as on their implication in a pathological process. This is an important step toward the elucidation of gene function, i.e. functional genomics, according to the French team who participated in this work – National Centre for Scientific Research Genexpress, Villejuif, near Paris.

But genomics should be backed up with proteomics, transcriptomics, glycomics (identifying the carbohydrate molecules, which often affect the way a protein works) and metabolomics (studying the metabolites that are processed by proteins). There is even bibliomics and bioinformatics which store and compare the sequences of genes and proteins, and mine the published scientific literature for discovering connections between all of the above. But as Sydney Brenner, the 2002 Nobel Laureate of Physiology and Medicine, once observed, in biotechnology, the one -omics that really counts is economics.

Most of the innovation in medical biotechnology, including the increasing reliance on genomics, has been done by small companies, the so-called start-ups, in close collaboration with the universities. The latter across the USA became hotbeds of innovation, as entrepreneurial professors took their inventions (and graduate students) off campus to set up companies of their own. Since 1980 (when the Bayh-Dole act was enacted), American universities have witnessed a tenfold increase in the patents they generate, spun off more than 2,200 firms to exploit research done in their laboratories, created 260,000 jobs in the process, and in 2002 contributed $40 billion annually to the US economy.

One should also underline the strong support provided by the public research institutions to the US biotechnology and the encouragement thus granted to the private sector for investing in this area. Within their long-standing policy aimed at ensuring the US preeminence in life sciences research and its applications, the National Institutes of Health, for instance, was distributing $27.9 billion to researchers and universities in 2004. This budget is increased by the contributions of other ministries such as the Departments of Defence, Interior and Agriculture. Such big public investment in basic research encourages private investors. Thus, in January 2004, despite the careful approach of investors who bore the brunt of the 2000 stock exchange drastic fall, Jazz Pharmaceuticals – a one-year-old start-up – succeeded in collecting $250 million from private investors. Also funding associated with research to combat bioterror has helped many biotechnology companies specialized in immunology to survive.

If new drugs are to be discovered, exploiting genomics or otherwise information is one of the most likely tracks to success. Some companies have understood this since the beginning, e.g. Incyte, founded in 1991, and Human Genome Sciences (HGS), set up in 1992, both using transcriptomics to see which genes are more or less active than normal in particular diseases. But HGS saw itself as a drug company, whereas Incyte was until recently a company that sold its discoveries to others. As a result, in 2003, HGS had ten candidate drugs in the pipeline, whereas Incyte had none.

The Icelandic company DeCODE Genetics is trying to use medical data of people to search for the genetic roots of disease. It has attracted controversy since July 2000, when bio-ethicists accused the firm of invading people's privacy and of not trying very hard to obtain people's consent before using their medical data. Three years later, the firm's methods were still seen shady, but DeCODE Genetics has found 15 genes implicated in 12 diseases, including the 'stroke gene'. The harmful form of this gene, which may cause plaque build-up in the arteries, is as much of a risk factor as smoking, hypertension and high cholesterol amount. Drugs to counter the gene are years away, and there is currently no way of knowing which form one has. But DeCODE Genetics' chief executive, Kari Stefansson, announced a screening test could be ready in 2005. Trying to connect genes to diseases and creating drug-discovery platforms, e.g. SNPs (single-nucleotide polymorphisms)/haplotype-based drug discovery projects, are the objective of, for instance, Perlegen (Mountain View, California) and Sequenon, based in San Diego, which are studying people's genomes only at the sites such as SNPs where variation is known to occur. Perlegen is using $100 million of its start-up capital to record the genomes of 50 persons.

Proteomics has been picked up by Myriad (Salt Lake City) which formed a collaboration venture with the Japanese electronics firm, Hitachi, and Oracle, a US database company, to identify all the human proteins and to study their interactions through expressing their genes in yeast cells.

Genaissance, another haplotype company, is trying to connect genes not with diseases but with existing drugs, through examining how people with different haplotypes respond to distinct treatments from the same symptoms, e.g. the individual response to statins which regulate the concentration of cholesterol in the blood – a $13 billion market in the USA alone (Pfizer, Inc.'s statin, Lipitor, is the most sold drug worldwide, and in August 2003 AstraZeneca had been authorized by the US Food and Drug Administration – FDA – to market its statin, Crestor, a formidable competitor to Lipitor). This kind of work may lead to 'personalized medicine', i.e. to identifying an individual's disease risk and knowing in advance which drugs to prescribe. It would also help drug companies to focus their clinical trials on those people whose haplotypes suggest they might actually be expected to benefit from a particular drug. This approach will reduce the very high cost of testing drugs and will probably increase the number of drugs approved, since they could be licensed only for those who would use them safely.

Presently, only about one molecule out of every ten subjected to clinical trials is licensed. This drop-out rate explains part of the high cost necessary to market a drug, between $500 and $800 million. Another research trend in medical biotechnology is to modify the activity of proteins through acting on their genes. For instance, Applied Molecular Evolution has been able to obtain an enzyme 250 times more effective than its natural progenitor at breaking down cocaine. Genencor is designing tumour-destroying proteins as well as proteins that will foster the immune system against viruses and cancers, just like vaccines do. Maxygen has produced more effective versions of interferons alpha and gamma, to be tested in people, and was developing proteins that would behave as vaccines against bowel cancer and dengue fever (*The Economist*, 2003*a*). X-ray cristallography of proteins is an efficient tool to unravel their structure and can contribute to the design of a new drug. Thus Viracept, devised by Agouron (part of Pfizer, Inc.) and Agenerase, developed by Vertex Pharmaceuticals of Cambridge, Mass., inhibit the HIV-protease. Relenza™, developed by Biota Holdings Limited of Melbourne, inhibits the neuraminidase of the influenza virus. Protein three-dimensional structure can also be deduced from its primary structure, i.e. the sequence of its aminoacids.

Vast computing power is needed to that end. Thus, IBM Blue Gene project is intended to solve protein-folding problem, because the foreseen petaflop machine could make a quadrillion calculations a second. It was considered an outstanding performance to have a machine running at a quarter of a petaflop by 2004.

## Current achievements and prospects

### Hepatitis C

Hepatitis C virus (HCV), which is spread mainly by contaminated blood, was not isolated and identified until 1989. In 1999, the most recent year for which global figures are available, HCV was believed to have infected some 170 million people worldwide.

Another 3 million are added every year. In most cases, the virus causes a chronic infection of the liver, which, over the course of several decades, can lead to severe forms of liver damage such as cirrhosis and fibrosis, as well as cancer. According to the World Health Organization (WHO), hepatitis C kills around 500,000 people a year. It is less deadly than AIDS/HIV, which claims more than 3 million lives annually. However, its higher prevalence (nowadays, some 42 million people are infected with HIV), longer incubation period and the absence of effective drugs, mean that it is potentially a more lethal epidemic.

Effective new treatments for hepatitis C are not easy to develop, due to the fact that the HCV is hard to grow in the laboratory and, until recently, the only animal 'model' of the human disease was the chimpanzee, a species that is impractical (and many would argue unethical) to use for industrial-scale research. However, new cell-culture systems and mouse models have opened the way to further drug development. The NS3 protease of the HCV is a target, and scientists at Schering-Plough Research Institute, in New Jersey, have begun clinical trials with an inhibition of this viral protease. Vertex Pharmaceuticals, a biotechnology company based in Cambridge, Massachusetts, has another anti-NS3 drug, VX-950, which blocks its target at least in mice; it could be tested in humans in 2004.

Other substances aim at inhibiting the binding of HCV to liver cells in the first stage of infection. Among these is a compound from XTL Pharmaceuticals, based in Rehovot, Israel, which has been tested on 25 chronic sufferers. The drug is a monoclonal antibody designed to block the HCV outer protein, called E2, which the virus needs to attach to its target cells. Roughly in three-quarters of patients who received the compound, viral levels dropped significantly, with no serious side-effects. As a result, XTL Pharmaceuticals was testing the drug in HCV-related liver-transplant patients, in whom it was hoped to prevent the infection of the transplanted organ by hidden reservoirs of the virus. The company expected the results of the trials before the end of 2004.

It is probable that a combination of drugs attacking the viral infection from different angles will be the most potent weapon. And as with AIDS, success in drug making will raise the thorny issue of access to effective drugs. Existing treatments, combining alphainterferon and ribavirin (an inhibitor of viral replication) already cost $20,000, which puts them beyond the reach of most of the world's infected in developing countries.

Future treatments, including a possible anti-HCV vaccine, may be more expensive and one has to find the money to pay for them when they arrive on the market.

## Ebola fever

Ebola virus is named after a tributary of the Congo river, close to the City of Yambuku (Zaire) where it was discovered in 1976 during an epidemics which affected 318 persons and killed 280. This is one of the longest viruses known to date; it is made of a nucleic acid thread embedded in a lipid capsid. The incubation period of the disease varies from a few days to three weeks and the symptoms include fever, intense abdomen pain and haemorrhagic diarrhoea with liver and kidney dysfunction. The virus is transmitted through direct contact with contaminated blood, saliva, vomiting, faeces and sperm; infected people should be put in quarantine. The haemorrhagic fever caused by the virus infection results in the death of 80% patients in a few days. Over the last few years, several of these fulminant epidemics have simultaneously occurred in the Democratic Republic of Congo and Gabon, thus placing the infection by Ebola virus as a major public health priority for these countries. It should be noted that the Ebola virus which causes havoc in Gabon and the Democratic Republic of Congo belongs to the most virulent of the four subgroups known, the Zaire subgroup.

Researchers for the French Research Institute for Development (IRD), associated with those of the International Centre of Medical Research in Franceville, Gabon, the World Health Organization (WHO, Geneva), the Wildlife Conservation Society (USA), the Programme for the Conservation and Rational Use of Forest Ecosystems in Central Africa (ECOFAC, a non-governmental organization in Gabon), the South-African National Institute for Communicable Diseases Control and the

US Center for Diseases Control in Atlanta, have been studying the virus since 2001 in the west of Central Africa. They assume that human epidemics caused by the virus originate from two successive waves of contaminations: a first wave of contamination moves from the virus reservoir to some sensitive species, such as mountain gorillas, chimpanzees and wild bovidae; then a second wave that infects humans through carcasses of animals killed by the virus. According to the epidemiological data collected during the human epidemics that occurred between 1976 and 2001, each epidemic evolved from a single animal source and then spread through the contacts between individuals. However, the study carried out between 2001 and 2003 in Gabon and the Democratic Republic of Congo suggests the existence of several distinct and concurrent epidemic chains, each one originating from a distinct animal source. In addition genetic analyses of the virus performed on patients' blood samples have shown these chains did not stem from a common viral strain but from several strains.

On the other hand, the counting of carcasses found in the forests and the calculation of the indices of the animals' presence (faeces, nests and prints) have revealed an important increase in mortality among some animal species before and during human epidemics. Gorilla and wild bovidae populations have been halved between 2002 and 2003 in the Lossi sanctuary (320 km$^2$) in the Congo, while the population of chimpanzees was decreased by 88%. Hundreds or even thousands of animals would have died during the last epidemics that occurred in the region. It was verified that the decline in animal populations was due to infections by Ebola virus. Genetic analysis of samples taken from the carcasses has shown the presence of several strains of the virus, like in humans.

In conclusion, epidemics caused by the Ebola virus among apes would not result from the propagation of a single epidemic from individual to individual, but rather from massive and simultaneous contaminations of these primates from the reservoir animal in peculiar environmental conditions. Human contamination occurs in a second stage, generally through the contact with animal carcasses. Consequently, the finding of infected carcasses can be interpreted as a sign of a forthcoming human epidemic. The detection, followed by the diagnosis of the infection by Ebola virus in animal carcasses, would allow the development of a prevention and control programme of the virus transmission to humans before any epidemic occurs; this would increase the probability of mitigating these epidemics or even avoiding them.

A vaccine against the Ebola virus, consisting of an adenovirus into which have been transferred the genes encoding proteins of the Ebola virus, has been made by the San Diego-based biotechnology company Vical. These viral proteins would induce the synthesis of antibodies against the Ebola virus in the persons infected. As the vaccine does not contain any virus-derived structure, it is theoretically harmless. On 18 November 2003, the US health authorities announced a first clinical trial aimed at studying the innocuity and efficiency of that vaccine. The US National Institutes of Health (NIH) indicated that the first phase of the clinical trial will involve 27 volunteers between 18 and 44 years. Among them six will receive a placebo and 21 the vaccine in the form of three injections over a two-months period. The volunteers will be under medical monitoring for a whole year. The clinical trial is a follow-up to experiments carried out on monkeys by Gary Nabel at the Vaccine Research Center of the National Institute for Allergies and Infectious Diseases. These experiments conducted for three years led to the complete immunization of the animals.

According to NIH director, Anthony Fauci, an efficient vaccine against Ebola fever/virus would not only protect the most exposed people in the countries where the disease naturally prevails, but would also be a deterrent weapon against those who might use the virus in bioterrorist attacks. In addition to the vaccinia virus and anthrax, US specialists who fight bioterror have been concerned for years about the possible use of pathogens causing haemorrhagic fevers, and particularly the Ebola virus.

## RNA viruses

### Human immunodeficiency virus

The human immunodeficiency virus (HIV) that causes AIDS (acquired immunodeficiency syndrome) shows a very great genetic variability and is particularly virulent, probably because of its recent introduction into human populations. Its evolution potential is very rapid, at the level of a population or an individual, due to its mutation rate among the highest in living beings and its capacity to recombine. Such potential is a major obstacle to the production of an efficient vaccine. Choisy *et al.* of the joint research unit of the French Research Institute for Development (IRD) / National Centre for Scientific Research (CNRS) / University of Montpellier II, devoted to the study of infectious diseases following evolutionary and ecological approaches, in collaboration with the University of California, San Diego, and the University of Manchester, United Kingdom, have tried the adaptative mechanisms of several HIV strains at he molecular level. They have studied and compared the evolution of three major genes of the HIV genome, *gag*, *pol* and *env,* in several subtypes of HIV.

Genes *gag,* which codes for the capsid proteins and *pol*,

which encodes the synthesis of the virus replication key enzymes, are very stable and conserved in all subtypes. By contrast, gene *env*, which codes for the proteins of the external envelope of the virus – the targets of the immune system – would contain sites that are selected positively.

Mutations of this gene have a selective advantage because they would result in the diversification of the expressed proteins – that would not be recognized by the antibodies. However, these proteins must conserve their vital function of adhesion of the virus particle to the membrane of host cells (CD4 cells of the immune system). This would mean that on gene *env* two opposed selection forces would operate, one toward conservation and the other toward diversification.

The French researchers have confirmed a recent theoretical model proposed by US scientists in 2003, i.e. the HIV uses very big complex sugar molecules to escape from the host's immune system. These sugars would create on the virus surface a kind of 'shield' that prevents the fixation of human antibodies, without hindering the role of the envelope proteins in sticking the virus to its host cell. This finding applies to all tested HIV subtypes. It could lead to the development of new drugs against HIV/AIDS and eventually to a candidate vaccine against all HIV strains. More research will be carried out to check the validity of these first results and make in-depth studies of the variability of SIV strains among primates, from which HIV strains have evolved.

## SARS virus

The SARS (Severe Acute Respiratory Syndrome) epidemic probably originated in the Guangdong province by early 2002 and spread to 28 countries. This disease is due to a coronavirus, the genome of which is made of 29,736 nucleotide pairs and sequence has been published on 13 April 2003 – an impressive achievement. Edison Liu and colleagues of the Genomics Institute in Singapore compared the genome sequences of coronaviruses isolated from five patients with those of viruses studied in Canada, USA and China. In a publication in *The Lancet* on 9 May 2003, the researchers concluded that the virus was relatively stable, compared with other RNA viruses. There were differences between the genome sequences of the viruses isolated from patients in Hong Kong and those of the viruses isolated from patients in Beijing and Guangdong. Such variation is useful in the study of the virus dissemination and epidemiological follow up.

Chinese scientists have published in January 2004 in the journal *Science* their analysis of the evolution of the SARS virus. Led by Guoping Zhao of the Chinese National Human Genome Center in Shanghai, a consortium of researchers in Guangdong Province, Shanghai and Hong Kong have shown that as the virus perfected its attack mechanism in humans, its potency soared. Early on, it was able to infect only 3% of people who came in contact with a patient; a few months later, the infectivity rate was 70%.

Based on virus samples taken from Chinese patients in the early, middle and late stages of the epidemic, the analysis showed the increasing infectivity of the virus during its transition to humans, as a result of evolution at the molecular level; it was therefore better to control the virus at a very early stage when the infection rate is lower. The Chinese researchers studied the SARS virus spike protein, which enables the virus to enter a cell. They found that the gene controlling the design of the spike protein mutated very rapidly in the early stages of the epidemic, thus producing many new versions of the spike protein. The new versions were maintained, an instance of positive selection pressure.

In the later stages of the epidemic, the sequence of the gene did not change, as if the spike protein had reached the perfect design for attacking human cells. The gene was under a negative selection pressure, meaning that any virus with a different version was discarded from the competition. The evolution of the spike protein from the animal host of the virus to acquiring the ability to attack human cells began in mid-November 2002 and was complete by the end of February 2003, a mere 15 weeks later. Another gene which plays a key role in the replication of the virus remained stable throughout the period when the virus was successfully switching from its animal to human host.

The study was praised by Kathryn Holmes, an expert on SARS-type viruses (coronaviruses) at the University of Colorado, for its speed, the foresight in saving specimens from the critical early stages of the outbreak and its epidemiological analysis at the molecular level. This expert stressed that this kind of evolution will occur in the future, referring to other pathogens that have moved from animal to human hosts. The SARS virus had probably infected humans many times before, but had failed to establish itself until 2002, when one of its constantly mutating versions succeeded in infecting humans.

On 16 May 2003 in *Science* (vol.300, pp.1062-1063), Rolf Hilgenfeld and his colleagues of the University of Biochemistry at the University of Lubeck, Germany, published the structure of a proteinase that plays a key role in the replication of the SARS virus, as a result of their x-ray diffraction studies. This proteinase is present in two strains of the coronavirus: one that causes SARS in humans and the other infects pigs. One day later, the biotechnology company Eidogen, Pasadena, California, also published the structure of this proteinase.

Starting from structural model developed by Hilgenfeld and his colleagues, the German researchers suggested that a proteinase inhibitor – AG7088 – which Pfizer, Inc., tried against the virus causing cold, could be a good starting point for designing inhibitors of the SARS virus proteinase.

According to the data provided by the World Health Organization (WHO), the SARS virus has infected more than 8,300 persons, out of whom more than 700 had died. Many researchers are of the opinion that the virus had remained latent in an animal species before infecting humans. Yuen Kwok-Yung, a microbiologist of the University of Hong Kong Centre for the Control and Prevention of Diseases in Shenzhen and his team focused their research on exotic animals sold as a food delicatessen in a market of Guangdong; they bought 25 animals belonging to eight different species; they isolated the coronavirus from six civets and found the virus in another two species. The virus isolated from the civets was almost identical to that isolated from patients suffering from SARS, the difference being that its sequence had 29 extra nucleotides. Later on, Henry Niman of Harvard University discovered that out of 20 patients only one, originating from Guangdong, was infected by a SARS virus that conserved the extra 29 nucleotides. This patient might be one of the first humans to be infected before the virus shed out the extra nucleotides. Peter Rottier of Utrecht University made the hypothesis that the loss of these extra nucleotides rendered the virus infectious to humans.

Other researchers have been more careful, e.g. the WHO's virologist Klaus Stohr stated that the animals which harbour the virus are not necessarily its reservoirs, and that a large number of animals should be screened before concluding that the civet is the virus reservoir. Also it is not clear why some persons harbour the virus without becoming ill and why children are not affected to a large extent by this disease. There are several methods to detect antibodies and viral material, but it has not been possible to identify the virus during the first days of infection, when eventually patients feel well but start disseminating the SARS virus.

Scientists of the US National Institutes of Health are working on the production of vaccines based on killed or attenuated viruses, proteins or DNA. On the other hand, some scientists are being concerned by the eventual worsening of the disease as a result of vaccination, due to the interaction with the immune system, as occurs with a vaccine against the coronavirus causing the feline infectious peritonitis. The US National Institute for Allergy and Infectious Diseases (NIAID) has allocated $420,000 for developing a vaccine against SARS, using an adenovirus as a vector.

Two biotechnology companies announced their intention to develop therapies against SARS. One of them, Medarex, Princeton, N.J., signed an agreement with the Massachusetts Biological Laboratories of the University of Massachusetts School of Medicine with a view to producing a therapeutic human monoclonal antibody. The other company is Genvec, which is collaborating with the NIAID in the vaccine development project. On the other hand, the company Combimatrix (part of Acacia Research, Newport Beach, California) has developed, two days after the publication of the SARS-virus genome sequence, a microarray chip of the virus that is dispatched free of charge to key research centres. In Germany, the Hamburg-based company Artus, which is collaborating with the Bernhard Nocht Institute of Tropical Medicine, has used the data of the virus genome sequence to develop the first diagnostic test, with a view to selecting targets for drugs against the virus.

## Avian flu virus

In 1997 in Hong Kong, bird or avian flu virus first demonstrated an unprecedented ability to infect humans. On the evidence of influenza specialists, the government ordered all 1.4 million of the territory's chickens and ducks to be slaughtered. Although 18 persons were infected, of whom six died, the swift culling eliminated a potentially much greater disaster. By early 2004, the avian flu virus strain H5N1 had infected mainly chickens – 80 millions of them had died from the flu and killed, burnt, or bagged or burried alive, in an effort to keep the disease from spreading. The virus had infected and killed 22 people up to February 2004. Most of them had probably come into contact with the birds' faeces or perhaps inhaled infected dust blown by flapping wings. Health officials were concerned less about the danger to farm workers than to the wider public. Should strain H5N1 acquire the ability to pass from human to human, instead only from bird to human, the consequences would be much more dramatic than the SARS epidemic.

The geographical scale of the 2004 avian flu outbreak is, of course, a critical reason why eradicating it is so much harder than in 1997. Flare-up of avian flu has been confirmed in 10 Asian countries and territories: China, Japan, South Korea, Taiwan, Thailand, Vietnam, Indonesia, Cambodia, Laos and Pakistan. Many scientists believe that migratory wild birds, which can carry many viruses without showing disease symptoms, were most likely the agents of the initial outbreak of the disease. Other factors, such as the transport of infected chickens across borders, legally and illegally, as well as of the reluctance and ambiguity of governments to acknowledge the reality of the outbreaks, came into play and caused the dramatic concerns.

Many analysts and media professionals have stressed the

official stonewalling and reluctance to acknowledge mistakes and to swiftly take the preventive measures. For instance, while the head of the Bogor Institute of Agriculture's faculty of animal husbandry suspected an outbreak of avian fly as early as August 2003, Indonesian officials finally admitted on 25 January 2004 that the country was facing a major outbreak of avian flu. In Thailand, the idea of a bird-flu epidemic was dismissed as an exaggeration that would damage the country's poultry exports (Thailand is the world's fourth-largest exporter of poultry), and harm farmers and workers involved in the chicken industry. Avian flu has been detected in nearly half of Thailand's 76 provinces, and almost 11 million birds have been culled across the country. China, the world's second-biggest poultry producer, has been the source of many of the major flu viruses to hit the world in the past 100 years, due to its vast population of chickens and ducks living in intimate proximity to each other and their human owners. After weeks of hesitation and ambiguous information, despite the former experience with the handling of the SARS epidemic, on 27 January 2004 the central government acknowledged the bird-flu outbreak had reached China. The ministry of agriculture, in a radical change toward an open approach to tackling the problem, demanded that all outbreaks be reported within 24 hours. By the end of January 2004, officials had confirmed outbreaks in Hunan and Hubei provinces in Central China, in addition to the cases reported across eastern China: Anhui and Guangdong provinces were potential hot spots, as well as Kangqiao, a suburb of Shanghai. The lesson to be drawn from both the SARS and avian flu epidemics is the following: when it comes to fighting highly contagious diseases, nothing is more important than decisive government intervention and transparency.

But stern government intervention is not enough to contain the bird flu. In Asia's countryside, almost everyone raises chickens or ducks; animals and humans live so closely together that the prospects of viruses spreading seem almost unavoidable.

Health and agriculture experts believe that livestock-husbandry practices are at the heart of the bird-flu crises, especially in South-East Asia. Changing these practices is a great challenge and the economic consequences of the epidemic should also be a warning for the countries to make the necessary changes. In this respect, while the US and Chinese economies are likely to expand strongly in 2004 – Merrill Lynch forecast Asian economies, excluding Japan's, will grow 6.1% in 2004, while US investment house T. Rowe Price expected corporate profits across Asia to surge about 15% – the Asian Development Bank's assistant chief economist warned if avian flu is not curtailed soon, it 'could cost the region tens of billions of dollars'. Thailand's $1.25 billion poultry industry was set to be devastated as exports to many markets were temporarily cut off. And tourism may also be threatened, although SARS epidemic had much more drastic consequences. In 2003 SARS-related estimated lost business revenue amounted to $59 billion in Asia (China: $17.9 billion; Hong Kong: $12 billion; Singapore: $8 billion; South Korea: $6.1 billion; Taiwan: $4.6 billion; Thailand: $4.5 billion; Malaysia: $3 billion; Indonesia: $1.9 billion; Philippines: $600 million; Vietnam: $400 million).

However, according to Daniel Lian, a Thailand analyst at Morgan Stanley, even if Thailand's poultry exports were to fall to zero for the first quarter of 2004, avian flu would reduce the country's total exports by only 0.4% in 2004. Unless the flu spreads, this analyst expects Thailand's projected 8% growth for 2004 to drop by only 0.11% of a percentage point. Other bankers and investors consider that the economic prospects for Asia look bright. Finally, the ultimate economic impact of the avian flu epidemic will depend not only on how quickly governments control the spread of the disease but also on how deftly they manage international perceptions of the threat.

The world has been afflicted by human flu epidemics, such as the 1918 Spanish flu pandemic which claimed up to 50 million lives. After Jonas Salk's efforts to improve influenza vaccines in the 1940s, 46,000 people died in the 1968 Hong Kong-flu pandemic. In 1976, swine-flu vaccine produced polio-like symptoms, and in 1997 avian flu claimed its first human victims, but it did not spread among people. The deadlines of the avian flu strain H5N1 drew the attention of scientists: the 1918 flu killed up to 4% of those infected, and SARS in 2003 killed 11%; in 1997, out of the 18 people infected with the H5N1 in Hong Kong, six died, i.e. the mortality rate was 33%. What really concerned the scientists was not only its mortality rate, but the persistence of the avian flu strain in trying to cross the species barrier. Sooner or later, they feared, it would infect a human and, through random mutation, adopt a form that allowed humanto- human transmission. Or, it could acquire this ability by swapping genes with another flu virus already adapted to humans. Or, the same event could occur in a pig infected with both the avian flu virus and a human flu virus.

There is no human vaccine for avian flu. Under the World Health Organization (WHO) aegis, laboratories in the USA and United Kingdom have begun developing a vaccine seed from viral specimens sampled from the 2004 outbreak, but this development and testing could take up half a year. Nine pharmaceutical companies make more than 90% of the influenza vaccine in the world. Diverting those resources to stockpile an avian-flu vaccine will take up time and will disrupt the supply of human-flu vaccine. The deadly nature of the bird-flu virus presents another drawback. Flu-vaccines are generally prepared from viruses cultured in fertilized hen eggs; but the H5N1 virus is as lethal to the embryo inside an egg as it is to adult birds. Vaccine developers have 'therefore' to use another vaccine production process, the socalled 'reverse genetics'. In 1992, Peter Palese of Mount Sinai Hospital in New York developed a technique for replicating RNA viruses through the replication of RNA into DNA and back again. The technique was refined by several research teams over the following years. Virologists Robert Webster and Richard Webbly from St. Jude Children's Research Hospital in Memphis, Tennessee, used this 'reverse genetics' technique to produce an experimental for the 1997 bird-flu strain H5N1. Firstly, they clipped off genetic material in order to make the virus no longer contagious. They then combined RNA from the virus with that of another known virus, called the 'master-seed virus', turned the whole lot into DNA, and tested it to make sure that the combination worked. They replicated the DNA back to RNA and obtained a genetically-engineered flu virus. This result did not raise much interest. Vaccine companies were concerned about licensing arrangements with MedImmune – the firm that holds the patents for 'reverse genetics', while regulators were wary of the genetic engineering approach, that could entail years for regulatory approval.

However, the rapid spread of the virus strain H5N1 has changed the situation. The WHO's Global Influenza Program director, Klaus Stohr, claimed a teleconference on 4 February 2004 with regulators and vaccine makers, who expressed their will to work together and solve the numerous logistical problems relating to large-scale vaccine distribution. Webster and Webby, after obtaining virus samples of the saliva of two Vietnamese patients infected with strain H5N1, were trying at St. Jude Children's Research Hospital to attach the genes coding for two surface proteins of H5N1 virus envelope to the master-seed virus, and thus develop a virus with the ability to trigger the human immune system. They expected to produce a vaccine within a few months.

Scientists are 'therefore' optimistic that a vaccine could be delivered before the bird-flu outbreak becomes a second deadline wave. If the pandemic extends before a vaccine is ready, physicians will have to rely on available drugs against standard flu. Unfortunately, strain H5N1 is already showing resistance to amantadine, a cheap and widely available drug. A more expensive drug, Tamiflu, will nevertheless remain effective against the viral strain, but supplies are limited, and it may be out of the reach of most Asian patients. It also seems that making a vaccine against flu, even a new strain of avian flu, is easier than creating one for an entirely new disease, as is required for AIDS/HIV and SARS. Although H5N1 has proved mutable, the WHO is confident that a vaccine will remain effective even if the virus undergoes a genetic shift that enables it to pass easily among humans. Yet, the immediate approach is to stop the dissemination of the virus among birds before a human vaccine becomes available; because, as stated by Yi Guan – a SARS and avian flu expert at the University of Hong Kong – 'once this virus can spread from human to human, region to region, it is too late'. By March 2004, avian flu had disappeared from the headlines, but it was not because the threat had receded. This risk of H5N1 avian flu strain mutating into a human transmissible strain remains as high as ever. Different strains of bird flu or influenza have spread to three countries with reliable animal health reporting systems – Japan, Canada and USA – but authorities were unable to stop it from spreading. On 3 April 2004, Canada reported that avian influenza had spread outside its quarantined 'hot zone' and infected at least two workers, with another ten suspected cases.

Before the H5N1 outbreaks were reported in Cambodia, China, Indonesia, Japan, Laos, Thailand and Vietnam, avian influenza had never infected so many birds over so large an area. Earlier outbreaks of less pathogenic avian influenza – Pennsylvania from 1883 to 1885 and Italy from 1999 to 2000 – were localized, but even so they required stringent controls over several years to extinguish. According to Klaus Stohr, the World Health Organization's Global Influenza Program director, 'given how far H5N1 avian influenza has spread, the world will be on the verge of a pandemic for at least a year, more likely two years'.

It is true that living in close proximity with animals exposes humans to animal viruses against which we have neither antibodies nor genetic immunity. Often, only one small mutation can change a virus' animal receptor into a human receptor. As a result, some of the biggest scourges have been diseases of animal origin. In the case of H5N1 strain, the pandemic risk arises if a person is simultaneously infected by both the avian influenza virus and by a human influenza virus. Both viruses could exchange genes and a recombinant avian-human virus is produced and is effectively transmissible from human to human. However, in the case of H5N1, according to the available data, all infected people contracted the virus from birds. There is no confirmed evidence of human-to-human transmission yet.

There are several phases in the emergence of a typical pandemic: firstly, the virus becomes endemic in its animal host; then it crosses the species barrier to humans; and finally it becomes efficient at spreading from person to person. Such possibility and danger grow with the rise in international trade and travel. In the case of SARS virus, analyses indicate that it moved from its original animal host to humans by mid- November 2002.

Only in early 2003, did it develop an efficient mechanism of spreading between people. Once this was achieved, the virus spread around the world within weeks: it spread to Toronto via Hong Kong and killed Canadians before medical authorities even knew there was a major problem. Airborne influenza viruses are much more easily spread than SARS, partly because still-healthy carriers shed large amounts of virus. That is why the basic containment measures – isolation and quarantine – that were very efficient in the case of SARS would not be as efficient if we deal with a recombined avian-human influenza virus.

Bonte-Friedheim and Ekdahl believe that governments should implement the following preparedness measures in order to protect their populations from a pandemic caused by a recombinant avian-human virus:

- Contingency planning for health-care systems; a new influenza strain would rapidly exhaust health-care systems' resources, so plans to activate back-up facilities and staff should be drawn up;
- Strengthening of influenza-vaccine programmes in order to create background protection as well as more capacity to mass-produce a new type of vaccine if needed;
- Stockpiling anti-influenza drugs; current anti-viral drugs may not cure all cases of a new H5N1 avian influenza strain, but they could limit the infection and can protect health-care workers;
- Compliance of surveillance in countries with H5N1 avian influenza with the World Health Organization's guidelines and procedures.

### Chagas' disease

Chagas' disease or American trypanosomiasis affects some 20 million persons in the tropical regions of Central and South

America, and is caused by the flagellated protozoan, *Trypanosoma cruzi*, transmitted to humans by Triatomidae insects.

Researchers of the French Research Institute for Development (IRD) Unit on Pathogenies of Trypanosomatidae as well as their collaborators of the National Institute for Health and Medical Research have studied the life-cycle of the blood parasite, its virulence and role in the infection, so as to identify possible ways of preventing and controlling the disease. In addition to luring the immune system of the host, *T. cruzi* is found during its life-cycle in the form of flagellated cells which circulate and multiply in the bloodstream, and of non-flagellated cells which are intracellular and give rise to the flagellated circulating cells. Both kinds of cells are able to secrete a protein, Tc52, which has enzymatic and immunosuppressive activities. This protein inhibits the production of interleukin-2 (IL-2) – a cytokine necessary to the proliferation of T-lymphocytes, and as such has an immunosuppressive action.

Mice have been immunized with this protein and 'thereafter' infected with *T. cruzi.* The result has been a decrease in the mortality rate during the acute stage of the disease, thus showing the protective effect of Tc52. When mutants of *T. cruzi* lacking the gene encoding Tc52 were used to infect mice, the latter could produce interleukin-2 normally while showing attenuated symptoms of the disease. The French researchers have identified the minimal sequence of the protein that causes the immunosuppressive effect. It is now possible to design chemotherapeutic strategies, i.e. the inhibition of the 4enzymatic activity of Tc52 by anti-parasitic drugs, or vaccination protocols against the protozoan. Research is being carried out in collaboration with the INSERM and the Laboratory of Immunology and Therapeutic Chemistry of the National Centre for Scientific Research (CNRS) on the identification of Tc52 receptors existing on macrophages and dendritic cells and on the synthesis of specific inhibitors of these receptors, thus paving the way to the development of new medicines against Chagas' disease.

## Type-1 diabetes

Type-1 diabetes is the result of the disorders in the immune system: for reasons not yet well understood a combination of genetic and environmental factors induces the destruction of insulin-producing islet cells (Langherans islets) in the pancreas. Inflammation plays a role in starting this process. Yousef El-Abed at the North Shore- Long Island Jewish Research Institute in Manhasset, New York, suggested that a protein known as macrophage migration inhibitory factor (MIF) might make a logical therapeutic target. MIF signals to a network of inflammatory pathways. By the end of March 2004, at a meeting of the American Chemical Society in Anaheim, California, Yousef El-Abed announced he had found a compound that might halt inflammation through cutting the communication between MIF and inflammatory pathways. This compound, named ISO-1, binds to a particular site on the surface of MIF and prevents it from carrying out its inflammatory properties, in the laboratory at least.

Experiments were 'thereafter' carried out with a group of laboratory mice treated with a chemical called streptozotocin or ST2, which caused them to experience the high booldsugar amounts that characterize diabetes. Of these animals, those not treated with ISO-1 all developed chemical-induced diabetes. But those who were treated with ISO-1, injected for ten days in the seventh or eighth week of their lives, did not. Al-Abed and his colleagues also investigated the drug's effectiveness on another group of mice that had been genetically engineered to develop type-1 diabetes. Of these, 20 out of 22 were protected. None of the treated mice suffered from any side-effects.

Yousef El-Abed thinks ISO-1 could be easily adapted into a human oral medicine within a few years. Test kits would have to be developed to find the children who could benefit most. The US researcher envisages a population-wide screen, like the infant heel-prick test which is used to detect phenylketonuria in newborns. Individuals with high amounts of diabetes-related antibodies would be monitored until the amounts indicating an impending attack on pancreas islet cells. At that point, the drug would be administrated. Costly though such screening would be, it could save $100 billion a year spent treating diabetes in the USA alone; while type-1 diabetes accounts for around 10% of diabetes sufferers, it takes around 40% of treatment costs.

## Auto-immune diseases

After talks with the US Food and Drug Administration (FDA), the Irish drugmaker Elan plc and the US biotechnology enterprise Biogen Idec expected to file for approval of their multiple sclerosis (MS) drug Antegren (natalizumab) one year earlier than foreseen – in mid-2004. If approved, the monoclonal antibody could come to market in 2005 rather than its previous market of 2006. Biogen Idec and Elan are collaborating equally on the development of natalizumab for MS, Crohn's disease and rheumatoid arthritis.

The humanized monoclonal antibody prevents immune cells to leave the bloodstream and to migrate into chronically-inflamed tissue by targeting the selective adhesion molecule (SAM).

The growing and hotly-contested MS drug market is estimated by some analysts to be worth up to $4 billion.

Currently, it is dominated by the German drugmaker Schering (Betaseron), Switzerland-based Serono SA (Rebif), Israeli Teva (Copaxone) and Biogen, Inc.'s own Avonex. Shares in Schering, Serono and Teva fell upon the news.

British Celltech Group plc and US biotechnology enterprise Biogen Idec have entered into a collaboration for the research, development and commercialization of antibodies against the CD40 ligand (CD40L) protein for the treatment of auto-immune diseases. The CD40L protein is a key regulator of antibody-mediated immune responses. Blocking the interaction between CD40L on T-cells and CD40 on B-cells has been shown to reduce excessive antibody production and may help restore normal immune response in patients with a variety of auto-immune-related conditions, including rheumatoid arthritis, multiple sclerosis and inflammatory bowel disease.

Celltech Group plc will be responsible for the identification and engineering of new high-affinity antibodies against CD40L and will bear all development costs until the end of phase-1 human safety testing. Following completion of phase 1, Biogen Idec has an option to co-invest in the ongoing development of products. In this case, the companies will jointly develop and commercialize products, and will share costs and profits.

Alternatively, if Biogen Idec does not exercise its option, Celltech may elect to take the programme forward independently, and continue to develop and market products on an exclusive, worldwide basis. Biogen Idec would then receive royalties based on sales achieved by Celltech.

### Struggle against cancers

In the fight against cancer, next to surgery, aggressive treatments, such as radiation and chemotherapy, were often considered thirty years ago the only options for survival. Unfortunately, they routinely did patients as much harm as good by killing off all dividing cells, cancerous or healthy. After a cancer is identified and a combination of drugs to combat it has been chosen, there is no guarantee that the drugs will work. That is because no two patients are alike. Subtle differences in their genetic make-up often determine how well a cancer drug will be tolerated and how quickly it will be broken down in the body. Some individuals produce enzymes that can neutralize the more toxic side-effects of anti-cancer drugs, while others either lack such enzymes or have genes that make them more sensitive to the drugs' adverse effects. Researchers at Massachusetts General Hospital (MGH), for instance, found that changes in the gene coding for an enzyme in DNA repair can mean the difference between breast-cancer patients who can tolerate chemotherapy and those with a twofold greater chance of experiencing a toxic reaction.

In 1970, a new field opened up: how a person's immune system can be coaxed into fighting cancer? Ronald Levy, now chief of the oncology division at Stanford University School of Medicine, began his training in cancer immunology, when he signed up for a two-year stint at the National Cancer Institute (NCI). Those were exciting times – President Richard Nixon had just declared his 'war on cancer', and funds poured into research projects. The idea was to make the connection from laboratory science to clinical medicine.

Two research teams in the United Kingdom made some breakthroughs. Each healthy Blymphocyte expresses a unique 'marker' protein on its surface, which it uses for binding antigens. When B-lymphocytes detect a foreign substance in the blood, they divide and churn out the same protein; this then binds to the antigen, tagging the invader for destruction. In B-cell lymphoma, one cell also divides endlessly, but this time without a purpose. In 1975, George and Freda Stevenson, of the University of Southampton, discovered that, in each individual patient, all malignant B-cells displayed the same 'marker' protein which could be used as a target. At the same time, César Milstein and Georges Köhler, of the Medical Research Council's Laboratory of Molecular Biology in Cambridge, worked out how to make monoclonal antibodies (they later received a Nobel prize); they fused two types of mouse cells – a tumour cell and a lymphocyte; the new hybrid cell had properties of both, it was not only immortal, but could also produce antibodies forever.

When R. Levy realized the implications, it became clear that he not only had a way to identify and target all cancerous B-cells in one patient, but would also have the ability to create a potent customized therapy that would exploit a B-cell's unique marker protein. By 1975, R. Levy was hired by Stanford University School of Medicine and eager to test his ideas in his own laboratory. After asking a few colleagues from around the world, including the Stevenson team, he settled on developing a patient-specific antibody able to latch on the target directly. Although R. Levy was able to repeat his findings, many other researchers were not. There were two difficulties: in some cases, the human immune system reacted strongly against the foreign mouse molecules and neutralized the antibodies; and many antibodies were not aimed at the right targets and had little effect. It would take more than a decade to sort out these problems.

The promise of antibodies (or 'magic bullets' as they were known at the time) boosted funding for biotechnology start-ups round the world, and helped R. Levy to co-found a company, Idec Pharmaceuticals, in 1985. But commercial success did not come easily and it took more than a decade before a modified version of the antibody was developed by Idec Pharmaceuticals and commercialized, though not customized for each patient. In 1997, the US Food and Drug Administration (FDA) approved Rituxan – the first monoclonal antibody for use against cancer and also the first lymphoma medication to hit the market in 20 years. It has since become a $1 billion drug.

Many in the field give R. Levy credit for laying the foundation for Rituxan's success, but he also never gave up trying to make 'a drug of one's own'. Since 1988, he has led about a dozen small studies, testing a patient-specific cancer vaccine. Rather than offering protection against the disease, its aim is therapeutic, for those already afflicted. At least half the 200 or so people who received the medicine showed an immune response, and many stayed in remission for longer than the two years that are typical following chemotherapy. In a few cases, tumour shrinkage was observed, lasting four to five years. Several patients have remained disease-free for years after receiving the treatment. But, even though the vaccine eventually proves successful in large randomized clinical trials, can it be manufactured cheaply enough to be cost-effective? R. Levy remains as determined as ever to find a solution. For one thing, vaccines are only half as much trouble to make as antibodies. Rather than manufacturing the antibody, scientists isolate the 'marker' protein and mix it with an immunity booster, hoping that the formula will coax the patient's own immune system against the cancerous cells.

Other researchers have carried out their own studies in the field. The closest to success looks like being Larry Kwak, now a principal investigator at National Cancer Institute's experimental transplant and immunology branch. After completing a series of small, successful studies, L. Kwak has conducted late-stage trials, usually the last step before receiving approval from the FDA. So far, no cancer vaccine – customized or otherwise – has reached the market. But for B-cell lymphoma patients, the wait may be coming to an end. Data from Genitope – a biotechnology company in Redwood City, California – trials were expected to be released by the end of 2004, and if the FDA deems the study a success, approval could follow a year or so later – a quarter of a century after R. Levy successfully tested a monoclonal antibody against B-cell lymphoma in a patient, who was 88 years old by the end of 2002.

The following anti-cancer drugs are being used:

- Gleevec, approved by the FDA against chronic myeloid leukaemia and gastrointestinal stromal tumours, blocks key signals that cancer cells use to drive growth;
- Erbitux, a monoclonal antibody approved by the FDA against advanced colon cancer, blocks the epidermal growth factor, a key protein cancer cells need to continue dividing; - Iressa (gefitinib), approved for treating lung cancer, blocks the epidermal growth factor; and
- Avastin, the first drug approved by the FDA that starves tumours by blocking the development of new blood vessels (angiogenesis). The following anti-cancer drugs were still in trials by mid-2004:
- Tarceva, also aimed at blocking the epidermal growth factor, was being tested for treatment of both advanced lung cancer and pancreatic cancer;
- SU 11248 blocks tumours by interfering with their ability to sprout blood vessels and by inhibiting enzymes essential for growth;
- BAY 43-9006, under study in kidney cancer, interferes with

a newly-discovered group of growth signals;

- Tarceva + Avastin, could block together distinct but critical growth pathways of cancer cells.

Most of the newly-approved drugs work in only 10% to 30% of patients, but in those patients tumours routinely shrink to less than half their size. The number of new drugs that have been approved is small, their cost is high (at least $20,000 per cycle), and progress is slow. The five-year survival rate for all cancers was 63% in 2003, up from 51% in 1975, according to the American Cancer Society. But it seems that most of that improvement was attributed to the effectiveness of anti-smoking campaigns, not to better drugs. However, the better that researchers know a cancer, the better their chances of defeating it. As we know more about the behaviour of tumour cells at molecular level, we are becoming convinced that one-drug-one-cancer approach is not sufficient. Just as we are using a multidrug approach to attack HIV at different stages of its lifecycle, so too are cancer specialists beginning to treat tumour cells with combinations of drugs that can weaken a growing cancer by chipping away at its life-support systems. In coming years, doctors will think not of breast and colon cancers, but rather of the signalling pathways the cancer cell is using, e.g. growth factors, angiogenesis factors, etc.

Chapter-3

# Biotechnology Applications

Biotechnology came into vogue almost simultaneous with the onset of human civilization. Early humans depended on plants and animals for food, clothing, shelter and fuel. The change from hunting and gathering as a way of life to an agarian lifestyle led to selective breeding of plants and animals. The first attempts at agriculture involved spreading of seeds on the ground and harvesting of the plants that grew from them.

Over time it was noticed by them that certain plants from the same species had superior characteristics. Some produced a greater yield; some were better tasting, while others were more resistant to adverse environmental conditions.

By selecting seeds from these desirable plants they were able to produce a greater amount of high quality food. Similar practices led to the development of many breeds of domestic animals. For example the domestic cow has been developed to be a docile animal yielding much more milk. The biological processes of microorganisms had been in use used for 6,000 years to make useful food products, such as bread and cheese, and to preserve dairy products and crops.

Early examples of biotechnology involved manipulating entire organisms. Today it is possible to manipulate organisms at the molecular level. During the 1960s and '70s our understanding of biology reached a point where began the use of the smallest parts of organisms, their cells and molecules, in addition to using whole organisms. The biological molecules most often used are nucleic acids, such as DNA, and proteins. Our concept of the gene has changed from that of particles or 'unit characters' to that of segments of the DNA molecule. Bioengineering involves the manipulation of specific genes. For example the gene for human growth hormone can be extracted and inserted into bacteria that will in turn manufacture the hormone.

Better meaning of the word biotechnology can be derived by simply changed singular noun to its plural form 'biotechnologies', because biotechnology is a collection of technologies using cells and biological molecules.

Cells and molecules are extraordinarily specific in their interactions. Because of this specificity, the tools and techniques of biotechnology are quite precise and are tailored to operate in known, predictable ways. As a result, the products of biotechnology will be better targeted to solving specific problems, generating lesser side effects and having fewer unintended consequences. Specific, precise and predictable are the words that best describe today's biotechnology.

## The technologies and their applications

Here are a few of the new biotechnologies that use cells and biological molecules and examples of their applications in medicine, agriculture and environmental management.

### Fermentation

This was one of the earliest applications of biotechnology. Early humans realized that the by-products from glucose breakdown in microbes (bacteria and yeast) could be used in a number of processes. The baking industry uses yeast as a leavening agent. Carbon dioxide produced causes bread dough to rise. Yeast also produces ethyl alcohol for the brewing of wine and beer. Bacteria produce lactic acid for making yogurt and acetic acid for making vinegar. New fermentation processes are being used to produce a wide variety of products including antibiotics, hormones and enzymes.

### Cell and tissue culture

Large numbers of plant and animal cells in laboratory situations, when placed in appropriate environmental conditions, with required nutrients, most cell types will multiply. This technology has applications in a number of situations. For example:

- Implanting a desirable gene into a single plant cell and then cloning hundreds of new plants from this cell. All plants cloned will have the desirable characteristics of the original cell.
- Producing animal cell cultures for drug testing. This technique replaces the need to test potentially harmful drugs on live animals.
- Culturing human tissue. This makes it possible to replace large areas of human tissue (example skin) destroyed by disease or accident.

### Genetic modification or recombinant DNA technology

Genetic modification technology is often referred to as recombinant DNA technology. The ability to modify genetic information provides a strong foundation for the biotechnology. In genetic modification, single genes whose functions are known are moved from one organism to another using recombinant DNA technology. A sample of techniques in recombinant DNA methods includes: gene isolation and amplification, site-directed mutagenesis, viral infection, and plasmid construction. Currently genetic modifications are used to produce high yielding, disease and pest resistant varieties of crops, new and safer vaccines and drugs and biodegradable plastics.

## Genetic engineering technology

The joining of genetic material from two different organisms or genetic recombination occurs naturally as part of reproduction. When humans started selective breeding they were manipulating the genetic material of the offspring. This practice was restricted to closely related species. Today, a single gene with known function can be removed form one organism and transferred to a totally different organism. Thus, by introducing new genetic instructions that would cause the cell to produce needed chemicals, or carry out useful processes, or give the organism desirable characteristics. For example, hemophiliacs benefit from this technology because genetically engineered bacteria are being used to produce large quantities of Factor VIII, a protein involved in the blood clotting process.

## Protein engineering technology

It is used in conjunction with genetic modification to improve existing proteins, usually enzymes, and to create proteins not found in nature. These new and improved proteins will encourage the development of ecologically sustainable industrial processes because they are renewable and biodegradable resources. The chemical, textiles, pharmaceutical, pulp and paper, food and feed, metal and minerals and energy industries have all benefitted from cleaner and more energy-efficient production made possible by incorporating biocatalysts into their production processes.

## Monoclonal antibody technology

It uses a type of immune system cell that makes proteins called antibodies. Antibodies exhibit extraordinary specificity. The specificity of antibodies makes them powerful tools for locating substances that occur in minuscule amounts and measuring them with great accuracy. A monoclonal antibody is a type of antibody produced from a single cell known as a hybridoma. All antibodies produced by the hybridoma are identical and bind to the same specific target in the same way. Monoclonal Antibody technology uses the specifity of antibodies in a variety of ways. They include treating various diseases, and detecting the presence of drugs, bacteria, viruses, abnormal cells, food contaminants and environmental pollutants.

## Biosensor technology

Biosensor technology couples biological method with microelectronics. A biosensor is composed of a biological component, such as a cell or antibody, linked to a tiny transducer. Biosensors are detecting devices that rely on the specificity of cells and molecules to identify and measure substances at extremely low concentrations. When the substance of interest collides with the biological component, the transducer produces a digital electronic signal proportional to the concentration of the substance. Biosensors are used to measure many blood components, to measure safely of food and measure environmental pollutants.

## DNA fingerprinting and diagnostic techniques

Different people have different DNA sequences (except identical twins). DNA is first extracted from a tissue sample. Various enzymes are then used to cut the DNA at specific sites, into segments of different lengths. Because different people have different DNA sequences they will give rise to different sets of DNA fragments. Gel electrophoresis is then used to separate the fragments, which moves according to their size. This pattern can then be compared to a crime scene sample fingerprint or other known source. This technique is also used in gene mapping (locating genes on the chromesome). It is possible to detect recessive genes in healthy people (carriers). Genetic counselors can advise these people about the risks of passing such genes on to their children.

## Biodegradation

Various microorganisms are vital to any ecosystem. They break down organic material and return it to the soil for recycling. Composting is, thus, one of the oldest examples of environmental biotechnology. Modern environmental biotechnology is making use of microorganisms and enzymes to clean up problems such as oil spills and toxic waste sites, and to purify sewage.

The biotechnology industry is the oldest industry on the earth that continues to grow in current times. The benefits from newer technologies include solving world food shortages and the elimination of many diseases. The prospects of gene therapy, fixing defective genes in humans, will open up an entirely new avenue for disease treatment and prevention. It is now possible to detect genetic defects and genetic predispositions even in unborn children. The application of bioremediation will see extensive growth as cheap and effective means for cleaner environment. The biotechnology industry is sure to play a major role in the future development of man. However, as with all other technologies there is also a dark side. The release of genetically engineered organisms may cause environmental imbalances. With the growth of biotechnology comes great responsibilities and ethical choices are to be appreciated.

CHAPTER-4

# Industrial Biotechnology

Industrial (or white) biotechnology, the use of biotechnology in industrial processes, is a subject which is rapidly gaining priority on the agenda of those in industry, politics, academia and NGOs. Why? Because white biotechnology offers enormous opportunities, not just for the economy, but equally to our environment and to our society. Recent studies from McKinsey and the Oko-Institute, as well as reports from the OECD, demonstrate the need to pay serious attention to building a European strategy for white biotechnology jointly among industry, politicians and scientists.

The USA is moving fast in this field. The current US administration has adopted the stimulation of white/industrial biotechnology as part of its governmental programme and allocated a substantial budget to draft a 'road map' to facilitate the development and implementation of the use of this form of biotechnology. We should not simply copy the US policy. Instead, Europe and individual Member States should use the rich potential that this continent offers in terms of knowledge, industrial activities and academic research institutes.

EuropaBio, the EU Industrial Association of Biotechnology, urges all white biotechnology stakeholders to jointly discuss the benefits of applying white/industrial biotechnology in Europe. We would like to see us identify and action some concrete steps towards making the use of white biotechnology really happen on a large scale.

In this paper, you will find our initial recommendations, both for the Netherlands and for Europe as a whole. To realise the proposed steps, political support at national and EU levels will be very important. It would be very welcome if the Dutch government were to use the Dutch EU Presidency in the second half of 2004 to draw up an action plan, and to present that to Europe. We would hope to see white/industrial biotechnology programmes being secured within the existing EU Framework Programme 6 and in the new Programme 7.

White biotechnology offers the rare opportunity to create a triple win for People, Planet and Profit. Let's not pass up this promising development for Europe, from which both our own and future generations can benefit.

Industrial biotechnology, also known as white or environmental biotechnology, is the application of nature's toolset to the production of bio-based chemicals, materials and fuels.

Current practice in industrial biotechnology demonstrates that the social (People), environmental (Planet) and economic (Profit) benefits of bio-based processes go hand in hand. Substantial reductions of 17-65% greenhouse gas emissions could be realized, and a more profound shift towards bio-based chemicals could potentially account for up to 20% of the global Kyoto target. The potential economic value of industrial biotechnology for the chemical industry alone is estimated to be € 11-22 billion per annum by 2010. As white biotechnology is making the industry more sustainable, it is expected that benefits will also seen across a range of critical society-based areas.

The Netherlands has a long tradition in biotechnology. Dutch-based life sciences companies have an overall yearly turnover of more than € 49 billion, invest € 950 million in research and development every year and employ 255,000 people. A substantial part of these life-sciences activities are devoted directly or indirectly to industrial biotechnology.

The Netherlands has the infrastructure and potential to become a leading player in industrial biotechnology. To further boost the developments in the Netherlands, it is proposed to pursue the following recommendations, among others, during the Dutch Presidency of the EU, which will take place from 1st July to 31 December 2004:

For The Netherlands the following priority setting is proposed for industrial biotechnology: focus on bio-based chemicals, and secondly on biomaterials, and finally on biofuels.

In line with these strategic choices, the Dutch government should take concrete steps to fully support a Dutch taskforce in order to:

- create a vision and roadmap on industrial biotechnology for the Netherlands underpinning the strategic choices in white biotechnology;
- benchmark the Netherlands with other OECD countries on the development of a biobased economy;
- propose special R&D programmes in order to fill the gaps in the industrial biotechnology R&D portfolio (e.g. systems biology, biomaterials);
- select and launch two or three demonstration projects;
- create a top Dutch institute on industrial biotechnology based on, or as a follow up of the existing industrial biotech R&D

initiatives, which should operate as a European centre of excellence for science and education; and

- facilitate stakeholders dialogue by promoting public awareness and support for industrial biotechnology.

In addition to these measures, it is proposed to create substantial (tax) incentives for all (new) start-ups, including those initiatives in white biotechnology, based on measures taken in France, Belgium or the UK.

Europe has considerable assets in the field of industrial biotechnology: for instance 70% of the world enzyme industry is European and a high level of knowledge in the field of food technology and fine chemistry is located in Europe. Moreover, there is a strong political and public sentiment to improve industrial sustainability in Europe (Gothenburg objectives) and the objective to become the most competitive and dynamic knowledge based economy in the world by 2010. For Europe the following recommendations are made:

It is proposed that during the Dutch Presidency of the EU The Netherlands actively participates in the EU technology platform on sustainable chemistry, where industrial biotechnology forms an integral but independent part.

The main tasks of this echnology platform would be to:

- define a European industrial biotechnology vision and road map on industrial biotechnology;
- conduct a benchmark between Europe and the US and Japan on the development of bio-based economy;
- secure an industrial biotechnology programme within the EU Framework Programme band 7;
- establish public-private-partnerships, whereby some of the

Dutch initiatives can serve as a model;

- launch selected demonstrations and information programmes to increase public awareness and support in white biotechnology;
- create a transparent and supportive regulatory framework;
- implement (tax) incentives for start-ups based on the French, Belgium or UK example, and;
- encourage competitive price for sugars within the EU.

White biotechnology has a lot to offer to our society, it is our challenge to develop and exploit that on time !

## Objective

The aim of this position paper is to outline the importance of industrial biotechnology for the Dutch and European society as well as its economy, and to propose concrete recommendations to boost the further development within the Netherlands and the European Union, among others, during the Dutch Presidency of the EU which will take place from 1st July to 31st December 2004.

Industrial biotechnology, also known as white or even environmental biotechnology, is the modern use and application of biotechnology for the sustainable production of biochemicals, biomaterials and biofuels from renewable resources, using living cells and/or their enzymes. This results generally in cleaner processes with minimum waste generation and energy use (see the following Figure1).

Industrial biotechnology can be differentiated from pharmaceutical (red) biotechnology or agricultural (green) biotechnology. 'Red' biotechnology is confined to the healthcare sector, whereas 'green' biotechnology is applied to the agro-food sector.

Industrial biotechnology is mainly based on fermentation technology and biocatalysis. In a contained environment, genetically modified or non-GM micro-organisms (e.g. yeast, fungi and bacteria) or cell lines from animal or human origin, are cultivated in closed bioreactors to produce a variety of goods. Likewise enzymes, which are derived from these (micro-) organisms, are applied to catalyse a conversion in order to generate the desired products.

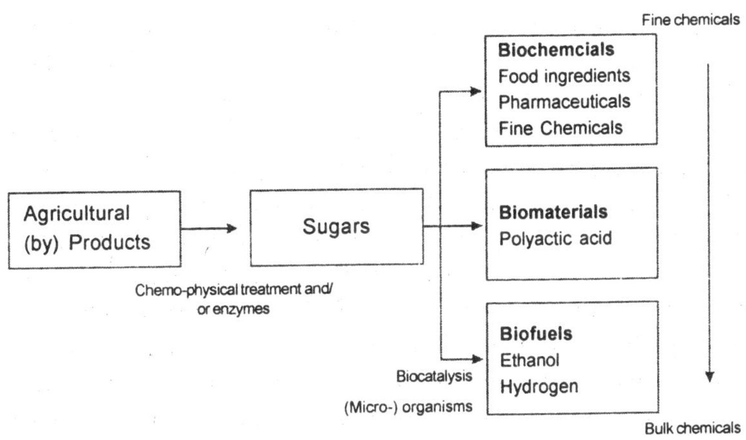


**Figure 1:** The industrial biotechnology value chain.

Activities and opportunities in this field are rapidly growing due to recent breakthroughs in genomics, molecular genetics, metabolic engineering, and catalysis. Promises are becoming reality and cells can now be used as tiny micro-factories, which can be optimised with respect to productivity, safety and minimal environmental load.

Figure 1 shows the industrial biotechnology value chain. Raw materials, including crops and organic byproducts from agricultural sources and households, are converted into sugars, which can be readily converted by tailor-made (micro-)organisms into the desired products. Typical products include enzymes, vitamins, flavours and fine chemicals such as chiral building blocks for the pharmaceutical industry. Traditionally, The Netherlands has a strong foothold in this value chain, given the presence of many important and international players in the agribusiness, food and chemical industry.

Today, the prime focus of the Dutch industry lies in the second part of the value chain, namely the fermentative and/or enzymatic production of biochemicals. This is the part where most value can be created.

The United States and Japan are Europe's major competitors in the field of industrial biotechnology. The US is strongly supporting industrial biotechnology1 and is spending nearly ten times as much as Europe on research in this field. Also China and other emerging countries are very fast developing in this field.

In the US, large R&D programmes are in place to convert complex (waste) biomass into sugars. Also the second part of the value chain is covered with emphasis on biofuels and biomaterials and to a lesser extent on biochemicals. The US 2020 vision1 is structured around a coherent strategy aimed at becoming less energy dependent. Whilst the orientation of the Japanese industry is not completely clear, most activities seem to be targeted towards the second part of the value chain (see Appendix). Several R&D programmes are in place and several new activities are planned.

The European Union distinguishes itself by a fragmented approach across the different Member States (see Appendix).

Every country has its own programmes and initiatives, with little or no EU-level coordination and no visibility in the EU Framework Programme 6.

Recently, a number of leading companies operating in industrial biotechnology, including DSM, Cargill Dow, Dupont, BASF, Novozymes and Genencor, in cooperation with the European and US biotechnology industry associations (EuropaBio and BIO) and the reputed and independent German Öko Institute, conducted an assessment of the potential impact of white biotechnology.

Detailed case studies were combined with a market analysis by McKinsey & Company to estimate the impact on the three elements of sustainable development: People, Planet and Profit (see Figure 2 below). The results confirmed an earlier study by the OECD that the social, environmental and economic benefits of industrial biotechnology go hand-in-hand. If all stakeholders cooperate in a self-reinforcing cycle, industrial biotechnology could create new employment, while reducing the impact on the environment and even creating economic value.

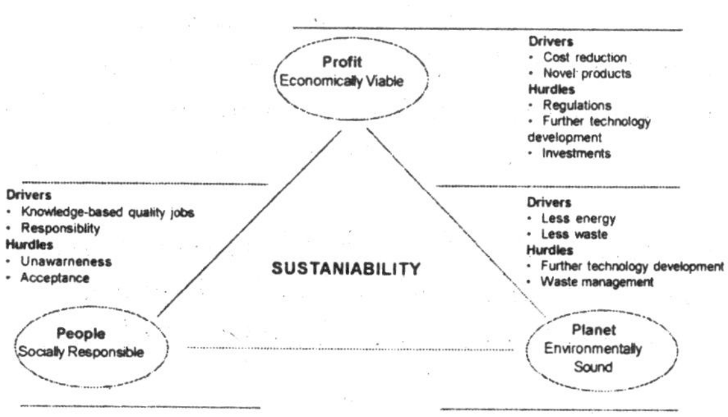


**Figure 2:** The triple-P bottom line.

McKinsey & Company estimate that biotechnology could be applied in the production of 10 to 20% of all chemicals sold by 2010, starting from the current level of about 5%. Whilst different chemical markets introduce and use biotechnology at different rates, the McKinsey study indicates that the greatest impact of industrial biotechnology will be on the fine chemical segment, where up to 60% of products may use biotechnology by 2010.

McKinsey estimates that between € 11 and € 22 billion additional added value could be created by the chemical industry alone in 2010, through cost reduction and the introduction of novel products. The economic impact of industrial biotechnology will, however, depend on the feedstock prices, technology developments and the policy framework as well as on the overall demand.

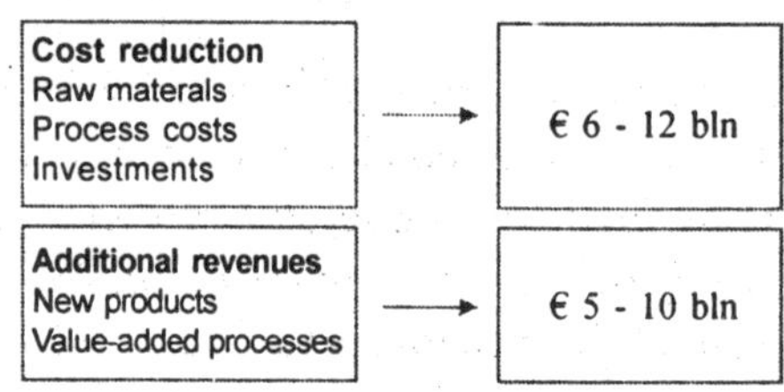


**Figure 3:** The impact of industrial biotechnology on the economy.

Starting with the chemical industry, white biotechnology will make inroads into a number of other industries. For example, enzymes will transform production processes in the pulp and paper industry and new polymers will find multiple applications in the automotive and consumer industries. Given the strength and presence of the Dutch biotech industry, it should be possible to capture a substantial share of this potential. According to the

Dutch Ministry of Economic Affairs, The Netherlands should be able to generate at least around €50 million extra economic value per year.

As industrial biotechnology moves from fine chemicals into the segment of commodities and eventually into bulk products, feedstock prices are becoming even a more important issue. The current price levels of renewable feedstocks for the fermentation industry exceed the prices of feedstocks used by the bulk and petrochemical industry. Moreover, current EU agricultural trade policy measures maintain the price of sugar at a level, which is higher than the world market prices. At present sugar costs € 596 euros per tonne in the EU, three times more than the world market price set at € 198 per tonne. Even further price reduction is necessary to allow the large-scale implementation of industrial biotechnology by developing cheaper processes to convert complex waste biomass as bagasse, cornstover and lignocellulose to simple feedstock as glucose, xylose and other sugars.

## Planet: environmental benefits

Industrial biotechnology is not an 'end-of-the-pipe' cleaning technology: it is a key tool in the development of sustainable production processes. As case studies have shown, industrial biotechnology has a substantial potential to reduce environmental impact: air and water pollution could be reduced, energy use lowered, fewer raw materials needed, and waste could be diminished or substituted by bio-degradable materials.

An environmental indicator that is relevant for all case studies on a global scale is greenhouse gas emissions. In their study, McKinsey & Company estimate that the application of industrial biotechnology in the chemical industry could considerably reduce global greenhouse gas emissions by 2010. According to

McKinsey there is sufficient agricultural byproducts in the world to satisfy at least 40% of the current demand for bulk chemicals. The shift to bio-based feedstock alone could potentially account for up to 20% of the global Kyoto target. This positions industrial biotechnology amongst the key technologies that can help to address global warming, one of the world's most pressing environmental challenges. Cleaner industrial biotechnology processes could thus enable countries to meet the Kyoto objectives in terms of carbon dioxide emissions.

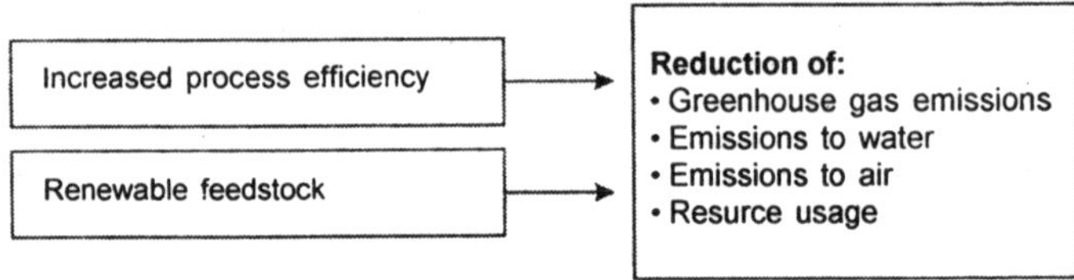


**Figure 4:** The impact of industrial biotechnology on the environment.

Industrial biotechnology has a dual impact: it increases process efficiency and enables the use of renewable feedstock. This entails a reduction of greenhouse gas emissions such as CO2, emissions to water, emissions to air and resource usage. The case studies performed by the Oko Institute have shown that reductions of 17 to 65% can be achieved for both fine chemicals and commodity chemicals. Even higher reductions have been estimated for future challenges, such as the bulk production of polyethylene from renewable resources.

During the fermentative production of biochemicals, biomaterials or biofuels, waste is generated in the form of microbial mass, which has to be disposed of once the product has been recovered. Typically, the microbial mass is inactivated by, for instance, a heat treatment and subsequently the mass can be incinerated or used as cattle feed or fertiliser (preferred in case of non-GM microorganisms having a GRAS status (generally recognised as safe).

Currently, the Dutch Ministry for Spatial Planning, Housing and the Environment (VROM) is conducting a life-cycle analysis and risk-assessment of several Dutch case studies to shape its policy with respect to industrial biotechnology.

## People: benefits for society

As industrial biotechnology makes industry more sustainable, it is expected that the benefits will be seen across a range of critical society-based areas: job retention/creation, development of new technology platforms, and the reduction of society's dependence on valuable fossil resources, thereby conserving them for future generations.

Industrial biotechnology can stimulate high-level education and research by providing highly qualified employment and by developing R&D initiatives such as the Kluyver Centre and B-Basic (see Appendix for details and further examples).

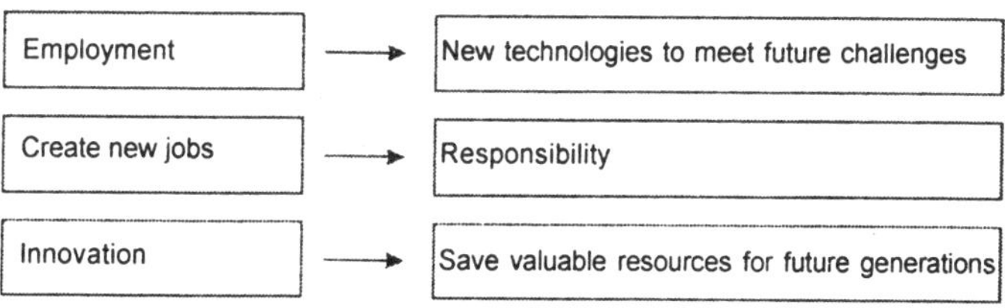


**Figure 5:** The impact of industrial biotechnology on society.

The Dutch Life Science Industry provides direct employment to more than 6,000 people in research and development alone. In addition to creating new and highly qualified jobs, industrial biotechnology can rebuild the industry by gradually replacing existing processes with a low technological level by highly specialised production processes and shifting to entirely new products. This will help to counter the unacceptable drain of highly qualified workers from Europe.

However, industrial biotechnology cannot achieve its full potential without a coordinated effort on the part of all stakeholders. As a first step, a dialogue amongst stakeholders needs to be started to share the facts and information as well as to discuss the opportunities, including the concerns related to this technology. Important stakeholders range from industry (fine chemicals, pharmaceutics, textile and leather, paper and pulp industry, recycling industry) to academia and public institutions. Stakeholders also include NGOs, the financial community, suppliers and industry users and observers from institutions such as the OECD. Currently, such stakeholder meetings are being organised in several member states and at the European level.

## Current obstacles to the development of industrial biotechnology

In Europe the rapid development of white biotechnology is hampered by a number of important obstacles. Key issues, which need to be addressed are:

- Development of a long-term strategy;
- Stimulating key technological capabilities;
- Setting favourable economical and regulatory framework conditions;
- Encourage competitive biological feedstock prices;
- Public awareness and acceptance.

Other countries have made considerable progress in addressing and breaking down these hurdles. In the US, for example, representatives from different governmental bodies, industry, agriculture, and academia worked together on a project called 'vision 2020' with the aim to boost industrial biotechnology usage over the next decade. Large R&D programmes are in place to develop improved and (much) cheaper enzymes for the conversion of agricultural (by)products into sugars (see figure 1). To compete successfully with the non-renewable - petro-based - bulk production processes, sugar prices need to be reduced further to about half of the current world market price. Likewise programmes are being started up to improve the next step in the industrial biotechnology value chain, the optimal conversion of sugars into valuable products by (micro-)organisms.

Finally, the Dutch government and the European commission could help to build broad public support for white biotechnology by increasing the awareness of its benefits along the three pillars of sustainability: society, environment and economy.

## The importance of industrial biotechnology for the netherlands

The Netherlands has a long history and reputation in both traditional and modern biotechnologya. The traditional industrial biotechnology includes the production of consumer goods such as food and beverages. The Netherlands also has a well established presence in modern industrial biotechnology, especially in the production of business-to-business biochemicals such as enzymes, antibiotics, food acids, flavours and other food ingredients. The overall sales of the Dutch fermentation industry amount to approximately 10 billion euros for 20036.

According to the Dutch Ministry of Economic Affairs, Dutch-based life sciences companies have an overall yearly turnover of more than € 49 billion, invest € 950 million in research and development every year and employ 255,000 people. A substantial part of these life sciences activities are devoted directly or indirectly to white biotechnology.

Although the Netherlands is an important world player in the production of high performance synthetic polymersb relatively little attention is being paid to the development of bio-based polymers. By combining our excellent R&D capabilities in synthetic polymers, as bundled in the Dutch Polymer Institute, with our strength in fermentation and biotechnology, a very strong biomaterials endeavour could be envisaged.

Besides biochemicals and biomaterials, industrial biotechnology also enables the production of bioethanol from renewable resources such as glucose (see Figure 1 above). Dutch companiesc are involved in the production of bioethanol. Bioethanol can efficiently replace products issued from the petrochemical industry. The European Union has agreed on a new objective: 5.75 % of all the fuel consumed by atmospheric engines should be biofuel. At present, the level is 0.3%. To meet this ambitious, but feasible objective, 9.3 million tons of bioethanol will have to be produced or imported per year by 2010. Public-private partnerships between industry, academia and institutes, such as the Kluyver Centre for Genomics of Industrial Fermentation and B-BASIC (Bio-based Sustainable Industrial Chemistry) have recently been set up (see Appendix) to further strengthen the Dutch R&D infrastructure in white biotechnology.

These initiatives are efficiently promoting pre-competitive research around themes of mutual interest. The prime focus of the Kluyver Centre is to optimise the microbial workhorses of the Dutch fermentation industry (e.g. baker's yeast, *Aspergillus niger* and lactic acid bacteria) by using genomic tools. Complementary to the Kluyver Centre is the recently approved B-BASIC initiative, which focuses on product and process development.

B-BASIC is part of the NWO programme Advanced Catalytic Technologies for Sustainability (ACTS), which also supports projects that combine chemistry and biocatalysis, thereby building on one of our key R&D assets.

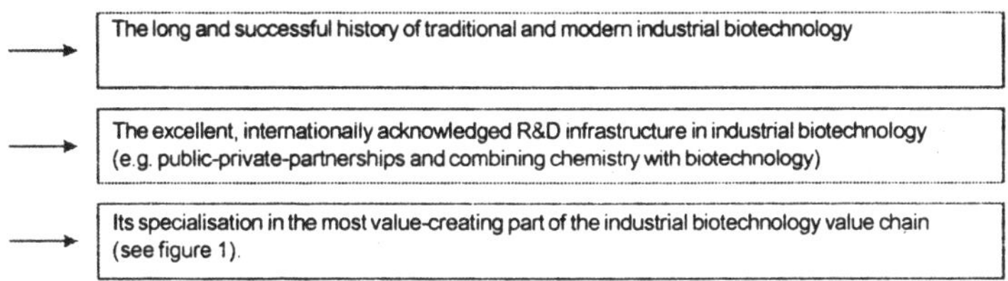


**Figure 6:** The three main assets of the Dutch industrial biotech.

Despite these activities some serious gaps in our knowledge base need to be bridged to fully capture the Dutch white biotech potential. A typical area, which urgently needs attention, is systems biology to completely understand and control microbial productivity. Furthermore, our efforts in biomaterials need to be increased substantially to keep pace with, and eventually outperform, the bio-based polymer initiatives in the US. To that end, a joined initiative of two of our public-private-partnerships (e.g., Dutch Polymer Institute and B-BASIC) would be highly recommended. Last but not least, demonstration projects could be started, where all relevant stakeholders are involved, with the aim to demonstrate the triple P benefits.

Recently, a Dutch taskforce for industrial biotechnology has been set up to create a consistent vision, coordinate our efforts and facilitate stakeholder dialogue. The taskforce is composed of representatives from the Dutch Ministry of Economic Affairs, the Kluyver Centre, B-basic, DSM and The Netherlands Genomics Initiative (NROG). Other parties will be invited to join when required. The ambition of the taskforce is to create a leading Dutch institute for industrial biotechnology, which is based on existing partnerships and which will be acknowledged as a world-leading centre of excellence for science and education.

At present about 13 start-up companies are directly or indirectly involved in white biotechnology. Further growth of existing as well as the creation of new ventures could be stimulated by appropriate tax incentives as recently introduced in France, Belgium and the UK. The Dutch authorities will soon be in a position where they can promote measures in favour of industrial biotechnology at the European level. Indeed, the Presidency of the Council is a privileged position to launch and support initiatives. Industrial biotechnology is a field in which major achievements can be reached for the food ingredients and chemical industry, with immediate results. The Dutch Presidency during their tenure between 1st July and 31st December 2004 should ensure that industrial biotechnology has the appropriate place on the European agenda.

## What should the netherlands do ?

The Netherlands has the infrastructure and potential to become a leading player in the field of industrial biotechnology, a strategic, knowledge-intensive field offering real economic, environmental and social benefits.

The main drivers for the Dutch industrial biotechnology are competitiveness (e.g., economic growth, high-quality jobs) and sustainability (e.g. less energy usage and waste production). These drivers differ from, for example, the US, which aims at reducing its energy dependency and at maintaining a high level of employment in agricultural areas.

For the Netherlands the following priority setting is proposed for industrial biotechnology:

- Biochemicals: build on current strength in (bio)pharmaceuticals, fine chemicals and food ingredients;
- Biomaterials: start up joint projects between B-basic and the Dutch Polymer Institute;
- Biofuels: focus on the production of bioethanol, either in The Netherlands from domestic waste, or from abroad, where cheap raw materials are available.

In line with these strategic choices, the Dutch government should take concrete steps to fully support the Dutch taskforce in order to:

- create a vision and roadmap on industrial biotechnology for the Netherlands underpinning the strategic choices in white biotechnology;
- benchmark the Netherlands with other OECD countries on the development of a biobased economy;
- propose special R&D programmes in order to fill the gaps in the industrial biotechnology R&D portfolio (e.g. systems biology, biomaterials);
- select and launch two or three demonstration projects;
- create a top Dutch institute on industrial biotechnology based on, or as a follow up of the existing industrial biotech R&D initiatives, which should operate as a European centre of excellence for science and education; and
- facilitate stakeholders dialogue by promoting public awareness and support for industrial biotechnology.

In addition to these measures, it is proposed to create substantial (tax) incentives for all (new) start-ups, including those initiatives in white biotechnology, based on one of the following measures:

- The French Young Innovative Company (YIC) status, which includes uncapped exemption of local business tax, exemption of social costs for employees involved in R&D for the first eight years, income tax exemption for the first three profitable years and 50% (or €100,000) relief of income tax for the following two years.
- The UK's new fund vehicle, called Enterprise Capital Fund, based on the US Small Business Innovation Company (SBIC) model. The main objective is to enable these Enterprise Capital Funds to use 'soft' government loans to leverage private capital and bridge the equity gap between business angels and private equity houses.

## The importance of industrial biotechnology for the european union

Europe has considerable assets in the field of industrial biotechnology: for instance 70% of the world enzyme industry is European and a high level of knowledge in the field of food technology and fine chemistry is located in Europe. Moreover, there is a strong political and public sentiment to improve industrial sustainability in Europe (Gothenburg objectives) and the objective to become the most competitive and dynamic knowledge based economy in the world by 2010 (Lisbon strategy).

Europe, however, invests less in R&D (1.9% of GDP in 2000 and even less after enlargement given that accessing countries have an average R&D level of 0.7%) than the US (2.7 % in

2000) and Japan (3% in 2000)2,11, and suffers from fragmented R&D funding and infrastructure. In the remaining six years, all Member States will have to take major steps to reach the Lisbon target of 3% by 2010.

A relatively new and helpful instrument to bringing Europe to the forefront of selected technology areas is a technology platform. A technology platform is a strategic, demand driven initiative, aimed at bringing together all interested stakeholders to address major economic, technological or societal challenges. The European commission has already scheduled the launch of three technology platforms, respectively for hydrogen and fuel cells, photovoltaics and water supply and sanitation technologies.

The setting up of a technology platform on sustainable chemistry, which would include industrial biotechnology, is currently being discussed. To facilitate this process, The Netherlands and Belgium have already organised themselves by installing a taskforce/platform on industrial biotechnology.

## What should the EU do ?

Industrial biotechnology is a promising sector, strategic and vital for European industry and the European economy. In addition to economic growth, the creation of highly qualified jobs and a reduced environmental load, the European Union can benefit from industrial biotechnology in terms of energy and farmland use.

The EU seems to cover the entire value-chain of industrial biotechnology (see Figure 1above). However, to fully capture Europe's potential a strategic vision and coordination is urgently needed. Clear choices must be made as to which particular areas of industrial biotechnology are to be supported (e.g., the production of sugars from biomass, biochemicals, biomaterials, and/or biofuels).

It is proposed that during the Dutch Presidency of the EU the Netherlands actively participates in the EU technology platform on sustainable chemistry, where industrial biotechnology forms an integral but independent part.

The main tasks of this technology platform would be to:

- define a European industrial biotechnology vision and road map on industrial biotechnology;
- conduct a benchmark between Europe and the US and Japan on the development of bio-based economy;
- secure an industrial biotechnology programme within the EU Framework Programme;
- establish public-private-partnerships, whereby some of the Dutch initiatives can serve as a model;
- launch selected demonstrations and information programmes to increase public awareness and support in white biotechnology;
- create a transparent and supportive regulatory framework;
- implement (tax) incentives for start-ups based on the French, Belgium or UK example, and;
- encourage competitive price for sugars within the EU.

Concrete and immediate action is now required to meet the challenges and seize the opportunities of industrial biotechnology.

CHAPTER-5

# Some Imperatives and Challenges for Rice Biotechnology

## Rice in Asia

Rice, *Oryza sativa L.*, is the staple food for more than three billion people or over half the world's population. It provides 27 per cent of the dietary energy supply and 20 per cent of dietary protein intake in the developing world. Grown in at least 114 mostly developing countries, rice is the dominant crop in Asia where it covers half of the arable land used for agriculture in many countries.Moreover, it is the primary source of income and employment for more than 100 million households in Asia and Africa. The Asian continent, where 56 per cent of humanity including 70 per cent of the world's 1.3 billion poor people live, produces and consumes around 92 per cent of the world's rice . Over 50 per cent of the 840 million people suffering from chronic hunger live in areas dependent on rice production. About 80 per cent of the world's rice is produced in small farms, primarily to meet family needs, and poor rural farmers account for 80 per cent of all rice producers. Nine of the top ten rice producing countries in 2003, viz., China, India, Indonesia, Bangladesh, Vietnam, Thailand, Myanmar, the Philippines, and Japan are in Asia. China and India combined account for more than half of the world's rice area and, along with Indonesia, consume more than three-fourths of the global rice production. However, less than seven per cent of the world's rice production is traded internationally. and, with this small marketable surplus, prices fluctuate widely with droughts, floods, and typhoons.

But more to being the world's most popular staple, rice has been cultivated by mankind for more than 10,000 years. Hence, it provides a symbol of global unity and cultural identity for many countries where rice cultivation is practically intertwined with religious observances, festivals, customs, folklore, and traditions. The United Nations, therefore, declared 2004 as the international year of rice, with a theme that captures the meaning of rice to so many people and cultures: Rice is Life.

## The challenge to increase rice productivity

The Green Revolution beginning in the 1960s ushered in an era of high rice productivity, with yields doubling or tripling from 1.9 tonnes per hectare (t/ha) in many Asian countries (Figure 1). Between 1966 and 2000, the rice production growth of 130 per cent, from 257 million Figure 1: Area harvested and yield levels (mt/ha) in major Asian rice-growing countries (FAO, 2003) tonnes in 1966 to 600 million tonnes in 2000, outpaced the population growth of 90 per cent in low income countries. The average per capita food availability was 18 per cent higher in 2000 than in 1966 (Khush 2004). About 84 per cent of rice production growth has been attributed to the use of modern technologies such as rice varieties that are semi-dwarf, early maturing, non-photoperiod sensitive and can, therefore, be planted more than once a year, and responsive to nitrogen fertilizer. More than 2,000 modern varieties, with resistance and tolerance to biotic and abiotic stresses, have been commercially released in 12 countries of South and South east Asia over the last 40 years. As a consequence, rice production cost per unit output was reduced by 20-30 per cent. This translated to reduced rice prices at the consumer level from about US$450 per metric tonne (mt) unmilled rice in the early 1950s to less than US$300 per mt by 1999 with world market rice prices decreasing by 80 per cent over the last 20 years (Cantrell and Hettel 2004).

In Asia, however, demand for rice is projected to increase by 70 per cent over the next 30 years, driven primarily by population growth that, excluding China, is expected to increase by 51 per cent. It is estimated that the Asian population will increase from 3.7 billion in 2000 to 4.6 billion in 2025. In the Philippines, for example, the population is expected to reach 107 million by 2025 and 65 per cent more rice, relative to present levels, would need to be produced to keep up with demand by a population continually growing by 2.3 per cent each year. This translates to a required sustained year on year rice production growth of 3 per cent. Given that annual rice production growth rates have been decelerating to less than 2 per cent per year, and the land frontier, the primary source of growth in recent years is closing, major technological progress has to be achieved in the next two decades to avoid importation. Against a backdrop of decreasing land, labour, and water and rising prices of production inputs, the challenge to further increase rice productivity is indeed enormous.

## Constraints in Asian rice production

Rice production and post-production processes in Asia are severely compromised by pests, diseases, and physiological and environmental factors. The rice crop, for example, is the world's single largest market for agrochemicals, consuming around US$3.7 billion annually, with agrochemical costs and crop losses amounting US$ tens of billions per year (DFID 2004). Tungro, the most destructive viral disease in South east Asia, for example, results in crop losses worth more than US $1.8 billion annually. Moreover, rice cultivation per se is restrained by resource constraints, among the most important of which are the projected scarcity of water, and scarcity of land. Technological progress is required to increase crop water productivity of rice and rice crop productivity under fragile environments such as the rainfed lowlands and uplands.

Biophysical constraints account for substantial yield losses in Asia. In the irrigated ecosystem, yield loss due to technical constraints accounted for 20 per cent (962 kg/ha) of the average yield, with oil-related problems being the most significant. On the other hand, yield loss due to technical constraints accounted for 33 per cent of average yield in the rainfed lowland and flood-prone ecosystems, with submergence being the most important, while it was more than 40 per cent of the average yield in the upland ecosystem, with drought being the most significant. Overall, all technical constraints caused a total yield loss of about 23 per cent or 833 kg/ha in Asia, with abiotic constraints being more important than biotic constraints for all ecosystems. Climate-related constraints like submergence, drought, and cold resulted in yield losses that ranged from 227 kg/ha (20 per cent of average yield) for the upland to 429 kg/ ha (28 per cent of average yield) for the flood-prone ecosystems. Yield losses due to pests and diseases, on the other hand, were most significant in the rainfed ecosystem while, that due to weeds were most important for the upland environment.

## The role for rice biotechnology

Agricultural biotechnology in Asia has been recognized as having the potential to: (i) increase crop and animal productivity; (ii) improve nutritional quality; (iii) broaden tolerance of crops for drought, salinity, and other abiotic stresses; and (iv) increase resistance of crops to pests and diseases. In the case of rice biotechnology, the Rockefeller Foundation (RF), in the process of developing its International Programme on Rice Biotechnology or IPRB identified the top-20 priority traits for biotechnology research intervention, balancing research costs vis a vis the benefits from expected increases in rice productivity or value.

Guided by this prioritization, the RF in the mid-1980s supported a 17- year programme that laid the scientific foundation for 'rice biotechnology' as we know it to-day. At about the same time, Asian National Agricultural Research and Extension Systems (NARES) began building biotechnology capacity. Among the important accomplishments of the IPRB were: (i) the generation of the first DNA molecular marker map of rice; (ii) the regeneration and transformation of rice; (iii) the use of rice pest genomic information to understand host-plant resistance; (iv) discoveries that changed the way rice geneticists viewed breeding objectives such as insect resistance, abiotic stress tolerance, and hybrid rice; (v) the discovery of rice's pivotal genomic position in the evolution of cereal species; (vi) the transfer of the resulting biotechnologies to institutions in rice-producing and - consuming countries; and (vii) the strengthening of both physical and human resources in cooperation with national and international rice research systems in Asia, Africa, and Latin America. The latter involved international collaborative research-cum-training that successfully linked emerging national rice biotechnology efforts directly to advanced research institutes in the United States, Europe, Japan, and Australia, resulting in the training of more than 400 rice scientists, primarily from Asia, in advanced laboratories around the world. At least 73 institutions in

12 Asian countries received both research grants while having up to 20 of their scientists sent to formal training activities that included: (i) Ph.D. fellowships; (ii) dissertation fellowships; (iii) postdoctoral fellowships; (iv) visiting scientist fellowships; (v) biotechnology career fellowships; and (vi) technology transfer fellowships in advanced laboratories and universities in developed countries.

Beginning 1993, the Asian Rice Biotechnology Network (ARBN) was also formed, with IRRI as coordinator, and this facilitated research collaborations amongst several Asian rice breeding programmes with a primary objective of developing disease resistant varieties through the application of DNA marker technology. Among the major Asian R&D institutions involved in the ARBN were the Indonesian Agricultural Biotechnology and Genetic Resources Institute, the Central Rice Research Institute in India, the Punjab Agricultural University also in India, the Philippine Rice Research Institute (PhilRice), the Agricultural Genetics Institute in Vietnam, and the China National Rice Research Institute (CNRRI).

## Progress in rice biotechnology applications

With increased activities on rice biotechnology beginning in the mid- 1980s, rice gradually became the 'model monocot plant' in molecular genomics research, eventually becoming the first food crop for which complete genome sequence became available. Progress achieved in the application of biotechnology for rice improvement has been in two major areas-the use of molecular markers for identifying and incorporating favourable genes within the rice species, and the use of transgenic technologies to incorporate traits for herbicide tolerance, biotic stress resistance, abiotic stress resistance, and nutritional value into rice.

**Use of molecular markers**—The development of the first rice molecular map in the late 1980s sped up molecular genetics research in rice. Among the early molecular marker applications for rice improvement were the: (i) construction of dense genetic maps using different populations; (ii) tagging and/or introgression of major genes and those underlying quantitative traits, referred to as quantitative trait loci (QTL); (iii) high-resolution characterization and fingerprinting of germplasm; (iv) assessment of the diversity of germplasm pools; and (v) map-based gene cloning.

Molecular markers offered great potential for increasing the precision and speed of rice breeding as, among other advantages over phenotypic markers, they provided the ability to (i) screen breeding populations regardless of growth stage; (ii) screen for traits that were extremely difficult, expensive, or time consuming to score phenotypically; and (iii) distinguish the heterozygous condition without need for progeny testing. Molecular markers provided geneticists with powerful tools to dissect the inheritance of economically important traits in rice, many of which are quantitatively inherited and complex in nature. Thus, studies dealing with QTL were carried out such as those involving tolerance to a variety of environmental stresses including drought, seedling vigour, submergence, salinity, and mineral deficiencies or toxicities. These traits were considered primary targets for molecular marker-aided selection (MAS) as breeding for them using conventional techniques often proved to be difficult.

MAS or the selection of traits based on the presence or absence of a molecular marker or markers in lieu of phenotype has already received a lot of emphasis in rice. The development of simple and less costly marker systems based on the polymerase chain reaction (PCR) such as the simple sequence repeats or SSRs contributed greatly to the use of MAS in various laboratories in developing countries. For example; at PhilRice, the NARES for rice in the Philippines, MAS studies are conducted to develop varieties resistant to bacterial blight, including the pyramiding of two to three bacterial blight resistance genes in a common genetic background, both for inbred and hybrid rice breeding. Gene pyramiding is expected to provide durable resistance against rice insect pests and diseases and attempts in this direction have been tried early on for bacterial blight and rice blast diseases, and the insect brown planthopper . Introgressing genes from wild relatives into cultivated rice has also been accomplished with the aid of molecular markers, such as the bacterial blight resistance gene from *O. longistaminata*, and the yield traits from *O. rufipogon.* Markers have also been used to minimize the linkage drag that occurs in wide crosses and to obtain the desired recombinants in fewer generations during backcrossing. Whole-genome, marker-based selection allows for new opportunities to unravel and makes efficient use of genetic variation both in cultivated rice and its wild relatives.

One of the most significant developments aided by the use of molecular markers in rice was the map-based cloning of *Xa 21* and its subsequent use in developing varieties with broad spectrum resistance to bacterial blight.Starting with the genetic mapping using RFLP makers of the *Xa 21* locus in 1990, the gene was cloned using map-based cloning techniques and a bacterial artificial chromosome library by 1995. By 1997, the gene had been pyramided with other *Xa* genes using PCR-based MAS and, by 1998, *Xa 21* had been transformed into elite lines with field trials conducted in China, India, and the Philippines by 1999.By 2000, a hybrid rice parental restorer line had been improved through MAS, resulting in resistant hybrid rices under field conditions.

The IRRI-coordinated Asian Rice Biotechnology Network that was supported by ADB and the RF played a key role in developing capacity for marker-aided analyses of pathogens and host plant resistance in several national breeding programmes. This network approach was found essential for the sharing of resources and providing sustained training in the adoption of new biotechnology tools and genetic knowledge in individual breeding programmes of different NARES in Asia. As a result of ARBN activities, elite or commercial rice lines with multiple disease resistance genes have been developed in several participating countries.

**Use of transgenic technologies**— No genetically modified (GM) or transgenic rice has yet been commercialized in Asian countries. However, two GM rice, both involving herbicide tolerance, have already passed regulatory approval processes in the US: the Liberty-Link™ rice of Aventis Crop Science (now Bayer CropScience) involving phosphinothricin (PPT) herbicide tolerance, specifically glufosinate ammonium, and the CLEARFIELD™ rice involving imidazolinone herbicide tolerance from BASF Inc. Ten trials in 11 hectares and 12 trials in 45 hectares were conducted in 2002 and early 2004, respectively, 90 per cent of which involved Monsanto.To indirectly gauge the extent to which the use of GM technology had so far advanced in rice, utilized information on patent applications and classified these into the areas of: (i) herbicide tolerance; (ii) biotic stress resistance; (iii) abiotic stress resistance; and (iv) nutritional traits. Until 2002, 307 patents on rice biotechnology from 404 different groups had been filed. The largest number of patents was held by DuPont/Pioneer (68), followed by Monsanto (33), Syngenta (32), Bayer (19), public sector institutions in Japan, and Japan Tobacco.

Amongst various traits, herbicide tolerance has been the major focus for the private sector. In the US, Monsanto and Bayer were responsible for 80 per cent of GM rice field trials, primarily addressing herbicide tolerance. Other countries where herbicide tolerant GM rice has been field tested include Italy, Brazil, Argentina, and Japan, and possibly China. Biotic stress resistance, on the other hand, has been the primary focus for public sector research institutions including those in Asia. Specific traits being worked on using GM technologies include resistance to bacterial blight using the gene *Xa*21, rice blast, rice hoja blanca virus, rice tungro spherical virus, rice yellow mottle virus, rice ragged stunt virus, the brown planthopper, and yellow stem borer, the latter, using *Bt* technologies, being the closest to commercialization. For abiotic stress tolerance, transgenic rice plants that produce trehalose at 3-10 times the normal rate, resulting in tolerance to drought and/or salinity have been developed by introducing the *ots A* and *ots B* genes for trehalose biosynthesis from *Escherichia coli* into rice. In China, Chinese researchers have developed several GM rice varieties that are resistant to the country's major rice pests and diseases, such as the stem borer, bacteria blight, rice blast fungus and rice dwarf virus. Significant progress has also been made with drought – and salt-tolerant varieties of GM rice, which have been in field trials since 1998.

Perhaps one of the most promising, despite controversies, application of transgenic technology in rice has been the development of Vitamin A-enriched rice, popularly known as Golden Rice™ due to the slight yellow color conferred on the rice endosperm in transgenic grains. Vitamin A is considered as absolutely essential for children and women of child bearing age and, worldwide, nearly 134 million children are at risk for diseases related to Vitamin A deficiency (VAD), including some

3.1 million pre school age children who suffer from eye damage, and nearly 2 million children under 5 years of age that die each year from diseases linked to persistent VAD. In Southeast Asia alone, 5 million children become at least partially blind every year due to VAD. Golden Rice™ has the potential to improve the supply of Vitamin A in the human diet, thereby alleviating the suffering and death of millions of people, especially those who cannot afford diet diversification (ISAAA 2004a). New Vitamin Aenriched materials with up to ten times more the level of pro-Vitamin A are now in the pipeline, including several popular Asian indica varieties such as IR64. Reported to involve "clean" events, without cross-border transfers or antibiotic markers, the new materials are being readied for backcrossing and stability and field testing in 2004, while vitamin A absorption and bioavailability tests are underway or planned in the Philippines, China, and the USA.

Another promising use of transgenic technology to improve human nutrition is in combating iron deficiency, one of the most widespread micronutrient deficiencies worldwide. Deficiency to iron in the human diet affects about 3.5 billion people worldwide and could result in illnesses such as anemia, heart problems, and neurological disorders. The ferritin gene from *Phaseolus vulgaris* has already been introduced into rice, resulting in the doubling to tripling of the iron content in the rice endosperm, even after polishing the grain. To improve the bioavailability of iron, since it is usually in complex with phytic acid, the genes from *Aspergillus fumigatus* encoding a thermotolerant phytase protein and the endogenous cysteine-rich metallothionein-like protein were also introduced into rice resulting in a 7-fold increase in cysteine level and a 130-fold increase in phytase level in the resulting transgenic plants.

Many NARES in rice-growing countries of Asia are actively involved in the use of transgenic technologies, encouraged by government policies supporting biotechnology research, and endowed with universities and agricultural research institutes with biotechnology research capacity. In a recent study done by the International Food Policy Research Institute (Atanassov *et al.* 2004), 209 transformation events were reported to have already been done in 76 scientific institutes in 16 countries. Of these, 109 (52 per cent) were done in 7 Asian countries, viz., China (30), Indonesia (24), India (21), the Philippines (17), Thailand (7), Pakistan (5), and Malaysia (5). Of these countries, however, only the Philippines had so far approved the commercial release of a transgenic food crop – a Bt-enhanced corn. Although the highest number of transformation events for any crop was reported for rice, followed by potatoes (11.0 per cent), maize (8.6 per cent), and papayas (6.2 per cent), a GM rice variety has yet to be commercialized in Asia.

In the Philippines, both IRRI and PhilRice have on-going rice biotechnology programmes employing molecular marker and transgenic technologies, as well as other more conventional techniques such as in vitro culture and wide hybridization. With the Philippine government declaring a policy supportive of biotechnology research, the use of biotechnology is embedded as a strategy for achieving the goals set by the irrigated lowland, direct seeded, rice for fragile environments, and hybrid rice multidisciplinary R&D programmes of PhilRice. GM technology, in particular, is being used to improve high-yielding varieties, including NPTs and hybrid rice parental lines. The Philippine focus is on the tungro, sheath blight, blast, and bacterial blight diseases, as well as on the insect stemborer, and tolerance to salinity. Genes procured from laboratories around the world and modified for *Agrobacterium tumefaciens*-mediated transformation, are being used to generate transgenic plants. A number of transgenic plants that contain chitinase and glucanase genes have already been produced and tested under controlled screenhouse conditions. Moreover, PhilRice has conducted the first and only contained field trials for any GM rice in the Philippines where the transgenic IR72 plants containing the *Xa*21 gene for bacterial blight resistance showed complete resistance against nine Philippine races of the pathogen. On the other hand, transgenic plants with the *pin*2 gene are being developed to improve stemborer resistance, while a coat protein gene from the rice tungro bacilliform virus is being used in *A. tumefaciens*-mediated transformation. PhilRice is also a member of the Golden Rice™ Network and undertook backcrossing activities on the now discarded original Golden Rice™ materials. To hasten the availability of vitamin-A enriched rice to consumers, PhilRice hopes to continue its active participation in this nctwork that involves other Asian countries such as Indonesia, Vietnam, India, Bangladesh, and China, as well as partners in developed countries such as the US, Germany, UK, and Switzerland. Already, guidelines are on the national testing of GM rice prior to commercialization are being prepared. To hasten public acceptance of biotechnology, in general, and GM rice, in particular, PhilRice, along with other Philippine government agencies, has also been spearheading a massive public education campaign using the tri-media as well as various public fora involving the government, private, NGO, and religious sectors.

**Work in progress**— Advances being made in the field of functional genomics will provide ample scope to further increase yield, build plant protection, improve nutrition, and enable rice to grow using less water and in adverse environment. The availability of the complete rice genomic sequence offers a lot of opportunities to further understand the natural genetic variation and the effects of alleles, and their interactions, for traits that are important in rice breeding, using specific genetic backgrounds and under specific environments. At IRRI, for example, biotechnologists are systematically assessing the array of phenotypes resulting from the disruption of putative gene sequences in mutants, near-isogenic lines, permanent mapping populations, and elite and conserved germplasm through a functional genomics initiative. However, these gene discovery and allele mining efforts would require the annotation of the rice genome and the subsequent build up of databases and information resources. The use of information and communication technology and bioinformatics, such as the IRIS (Bruskiewich *et al.* 2003; http://www.icis.cgiar.org/) and GeneFlow (www.geneflow.com) databases, should make the mounting information more easily accessible to scientists, especially breeders in rice-growing countries.

Other work in progress include studies designed to transfer the C4 photosynthetic pathway and leaf anatomy genes of maize to C3 rice in order to improve the rice plant's radiation use efficiency while reducing transpirational water loss and N fertilizer requirement, and studies aiming to more deeply understand the genetic variation for drought tolerance using genomics and bioinformatics tools in order to identify the exact genes involved. Another promising area is the genetically engineering of N2 fixation capacity into rice and, in this direction, attempts to engineer the nif-regulon into the chloroplast genome have been made with the idea of not having to make rice totally independent of external nitrogen supply, but to provide additional nitrogen during the grain filling period to rescue the photosynthetic apparatus for a longer period of time. Over the medium- and long-term, the apomixis research that earlier had been started at IRRI, would need to be vigorously pursued using biotechnology in order to capture and made available to resource poor farmers the benefits of heterosis.

With over 27 million tonnes of rice worth US$5560 million lost annually due to pests such as the stem borer, sheath blight, and bacterial blight, GM rice solutions are nearing commercialization to help offset this yield loss. Ready for field evaluation are Bt/hybrid rice in China and India; XA rice in China, Philippines and India; Golden rice in the US, the Philippines and India; and herbicide tolerant rice in Spain, USA, and China. For commercialization by 2005-06 will be Bt rice in China, and Golden Rice and Ferritin or high- iron rice by 2007-08.

## Issues, Concerns, and Opportunities

**Explicit governmental biotechnology agenda**— There is a need for many Asian rice-growing countries to develop clear and time-bound national agenda for biotechnology R&D and commercialization. For example, in the Philippines, a technology explicit and market driven national development agenda, which recognizes the role of science and technology in promoting economic development and facilitating trade is desired, along with corresponding increased investments in R&D. This could be achieved by developing a collaborative scheme that shall bridge the academy and science and technology community with industry, and having both well tuned to market demands. There is also a need to emphasize the role of the government in formulating technology policies and plans, as well as funding of research and development projects; the role of academy in identifying what problems need solving, and the role of the private sector in investing research that could meet the country's immediate needs.

In the case of India, for the first time, a specific impetus has been given to the advancement of biotechnology. In the country's 2005-05 budget, the Minister for Finance emphasized that science and technology, including biotechnology, will receive priority and will be provided with additional funds. A specific provision was given for companies doing scientific R&D approved by the Department of Scientific and Industrial Research before April 1, 2004 to be entitled to 100 per cent deduction of profits for 10 years. The Federation of Indian Chambers of Commerce and Industry noted that this will favour small and medium biotechnology companies who can channel savings to augment R&D activities.

The Agricultural and Development Economics Division (ESA) of the Economic and Social Department of the United Nations' Food and Agricultural Organization (FAO) showed that most of the research institutions in China, Cambodia, and Indonesia are in the initial states of developing biotechnology research capacity. The proportion of biotechnology expenditures to the total agricultural research expenditures is very small, and mainly from the public sector. The report discusses the importance of secured and sustained public research capacity including physical, human and financial, in the successful development of biotechnology innovations. For instance, countries like China, India, Indonesia, Malaysia, Philippines, Argentina, Brazil and Bulgaria have made significant investments in biotechnology research and regulation. Experience with genetically modified organism (GMO) testing for a wide array of locally important traits, and commercialization in these countries is growing, and government support programmes and policies actively encourage biotechnology R&D.

In China, scientists have released a report urging the central government to allow the commercial planting of genetically modified (GM) rice. They note that GM rice technologies are technically mature and are ready to be commercialized. GM products include several rice varieties resistant to China's major rice pests, including those that can resist the stemborer by using *Bacillus thuringiensis* (Bt), delta endotoxin and cowpea trypsin inhibitor CpTI genes, a protease inhibitor rice, a planthopper and bacterial leaf blight resistant rice using the Xa21 gene, and fungus-resistant rice. The country currently has the largest field for GM rice trials. An estimate of about 25-30 per cent of China's plant biotechnology investments are spent on GM rice programmes. China has increased its budget for research and field trials of GM rice since 2001. Its biotech budget for 2001-2005 is $1.2 billion, a 400 per cent increase compared with 1996- 2000. About $120 million out of the current budget is devoted to GM rice programmes. The Chinese Government has become the world's second-largest spender on plant biotechnologies, next only to the United States and the country is expected to launch at least 10 GM rice field trials by 2005 and release *Bacillus thuringensis* (Bt), cowpea trypsin inhibitor gene, and *Xa21* gene GM rices that year. Already, field trials in Hunan and Fujian provinces showed that GM rice boosted yields by 4 to 8 per cent, and allowed an 80 per cent reduction in pesticide use.

**Setting the biotechnology R&D priorities**— It may be argued that most of the earlier rice biotechnology activities, particularly in the public sector and on transgenic technology applications, were more science driven and researcher-driven than attuned to the needs of ordinary farmers. It is important to note that for GM rices to be useful, at least in the immediate term, and to gain fast acceptance amongst resource poor farmers, it is best that they be derived from varieties already widely grown and suited to specific agro environment. Ranged against the challenges confronting rice cultivation in most Asian rice growing countries today, it is clear that international as well as NARES biotechnology research should focus, on one hand, on the most important rice diseases and pests, and physiological and environmental factors that reduce rice productivity and quality as discussed earlier, and on the other hand, on increasing rice yield potential. An example of trait prioritization for rice biotechnology research in Asia, aimed at delivering the greatest impact on the lives of poor rice farmers and consumers, while being complementary to conventional rice improvement efforts, has been put forward by Hossain *et al.* To ensure relevancy of biotechnology R&D agenda, a bottom up approach needs to be implemented in the crafting of R&D priorities, where farmers' and other stakeholders' needs and concerns are adequately addressed. Such approach should benefit from the rich indigenous knowledge of local farming communities on specific rice production constraints while facilitating public acceptance and ensuring the trickling down of benefits from biotechnology-derived products.

The extent to which biotechnology may improve food security does not settle the question of its relevance to quality of life; it is essential to move stakeholder involvement upstream in setting research priorities and to fairly share benefits through appropriate infrastructure and ownership arrangements.

**Need for more public investments**— There remains an imbalance on R&D investments on rice biotechnology that tend to favour developed countries, thus impacting on the potential of biotechnology to boost agriculture in the developing world and alleviate the plight of resource-poor rice farmers in Asia. The concentration of biotechnology R&D in developed countries and the limited private sector effort in developing countries, particularly in Asia, has raised concerns over the economic concentration of biotechnology in favour of developed countries and multinational companies. As the private sector would be unlikely to undertake rice biotechnology research based primarily on the pressing needs of resource poor farmers, due to difficulty in recovering costly investments, there is need for significant publicsector funding initiatives if benefits are to reach resource-poor farmers in developing countries in order for them to develop pro-poor biotechnology R&D agenda. Furthermore, public research products would have to gain similar approval as those developed by the private sector if transgenic research products and their associated potential benefits are to reach the poor. In this context, it will be necessary to work with local communities to ensure acceptance and adoption from the bottom up rather than simply again trying to impose viewpoints from the top down. Thus, needs must be identified – nutritional and environmental– so that product traits are country-relevant.

While Asian scientists have demonstrated the capacity to successfully undertake biotechnology R&D relevant to the needs of resource-poor farmers, the desired phenotypes have been few when compared to traits being developed by multinational firms and advanced research institutes in the developed world. One bright spot, however, has been the case of Thailand that established the National Center for Genetic Engineering and Biotechnology (BIOTEC) in 1983. BIOTEC has supported biotechnology

R&D in six areas, including the improvement of disease resistance in rice, particularly rice blast. This disease affected 200,000 hectares of rice in Thailand in 1993, causing serious economic loss and resulting in government intervention to assist disease-struck farmers worth about US$10 million. Since then, BIOTEC has supported research for the molecular genetic characterization of local blast isolates and mapping of resistance blast genes, with focus on aromatic varieties for Thailand's export rice market. In 1999, BIOTEC also provided US$3.7 million to fund the 'Rice Genome Project Thailand', particularly for the sequencing of rice chromosome 9 that contains a QTL for tolerance to a very important concern of Thai rice farmers-submergence. Most rice growing countries in Asia, with the exception of China and India, however, have yet to launch similarly focused government initiatives on rice biotechnology R&D. In China, investments on public sector biotechnology research has risen dramatically to $1.2 billion for 2001-2005, a 400 per cent increase over 1996-2000 levels, with about $120 million allocated for transgenic rice R&D. With field testing on various transgenic rices already going-on since 1998, with 53 hectares planted in 2003, China is poised to becoming the first country in the world to commercialize transgenic rice.

**Importance of collaborations**—Given the varying biotechnology research capacities of rice-growing NARES and the limited resources allocated for biotechnology research in the public sector, unintentionally encouraged by the phasing out of the IPRB of the Rockefeller Foundation, the constraints in NARES R&D budgetary allocations, and the reduction of funding support for international agricultural research centers (IARCs) such as IRRI, the need for biotechnology R&D practitioners to collaborate have become paramount. Collaborations need to be pursued at the individual, institutional, governmental, bilateral, regional, and international levels to ensure not only that the highest returns for R&D investments are attained, but also to facilitate regulatory approvals and biotechnology product commercialization. At the national level, the creation of a national coordinating body such as BIOTEC in Thailand should provide a mechanism for increasing efficiency in the use of limited national

R&D budgetary allocations through the avoidance of research duplications and sharing of in-country research capacities. In the Philippines, the recent creation of a Crops Biotechnology Center under the Department of Agriculture augurs well for the development of a unified biotechnology R&D agenda for the country's priority crops, including rice, and the optimization of the use of limited manpower and logistical resources. On the other hand, a regional collaboration approach, as exemplified by the ARBN should be able to develop a biotechnology R&D agenda focused on the shared needs of rice farmers in the region and, where possible, human pool, scientific, and financial resources or, alternatively, parcel out the research portfolio as was done in the rice genome sequencing initiative. One type of formal collaboration that has yet to be explored involves bilateral arrangements between countries. In the development of transgenic technologies, these South-to-South collaborations would be facilitative to the build-up and sharing of common approaches, genes, germplasm, regulatory trials, and biosafetyrelated information. The already established broad-based regional cooperation, such as the Association of Southeast Asian Nations (ASEAN) and the Asia Pacific Economic Conference (APEC) should be tapped to support these regional biotechnology undertakings. At the international level, programmes that help rice scientists from developing countries to train, further hone their capacities, and maintain ties with advanced laboratories at IARCs and developed countries need to be supported. For instance, India's Department of Biotechnology (DPT) and the United States for International Development have signed a letter of intent to initiate enhanced cooperation in agricultural biotechnology research and development. The partnership will pursue agri-biotech research projects including technology development, technology diffusion, biosafety, and related policy activities. In addition, joint workshops, conferences, scientific exchanges, and training of scientists will also be done. The programme will increase the 'range of safe and environmentally sound technological options to producers and consumers of agricultural products'

**Role of IARCs and the private sector**— Without formal, dynamic, and synergistic interfaces between the public and private sectors, much of the benefits of crop biotechnology will not reach those who need them the most. The sharing of information and experiences across sectors is crucial to facilitate the flow and process that technologies undergo from the laboratory to the farm. With many NARES still not fully enabled to undertake, solely by themselves, activities spanning the whole biotechnology research, development, and commercialization spectrum, IRRI and similar international institutions will still need to play their roles as technology and knowledge providers, as well as builders and enhancers of biotechnology capacities of rice-growing NARES. Of particular importance for IRRI is the provision of strategic research outputs that already several NARES in Asian countries are capable of transforming into applications and products. These could include protocols, gene constructs, and markers for traits relevant to local problems but prohibitively expensive for NARES to develop by themselves. Alternatively, IRRI should be able to complement its strategic research programme with a product development thrust, focusing on biotechnology-derived advanced breeding lines and varieties, with traits that have common high relevance amongst Asian countries. The product development portfolio could include varieties tolerant to drought, of high nutritional value, and resistant to major diseases such as tungro and bacterial blight. The role of IRRI as facilitator in the transfer of useful technologies and products amongst NARES through the sharing of technologies, knowledge, and experiences need to be strengthened. Equally important is its role in facilitating the formation of effective NARES/public sector and private sector collaborations, for NARES to be able to access private sector-held intellectual property (IP) on rice biotechnology processes and products. IRRI can also serve as a clearing house for IP-protected technologies from both the public and private sectors that NARES can easily access. Training support by IRRI and similar institutions for NARES should now also include those designed to advance NARES capacity on the science and management of biotechnology, IPR, biosafety and food safety regulation, and international negotiations. As Cantrell and Hettel argued, with IRRI's strengths, it can serve as the unbiased broker and facilitator amongst the rice NARES, advanced research institutions, and the private sector.

Other international organizations, such as the FAO, can help expedite rice biotechnology progress in Asia by promoting and supporting networking mechanisms such as the South-to-South cooperation model. They can also help in developing and supporting infrastructure for public-good agricultural research, providing knowledge and training to NARES researchers, enabling interactions amongst biotechnology stakeholders through dialogues and similar fora, facilitating access to relevant IP, sensitizing policymakers on biotechnology-related issues, and assisting governments in the crafting of biotechnology-related policies. As the primary source of GM crops continues to be the private sector, technology transfer between private and public sectors, in terms of products as well as experiences on regulation, commercial development, and release of GM crops, would greatly benefit NARES. This technology transfer can be facilitated by private foundations such as the International Service for the Acquisition of Agri-Biotech Applications (ISAAA).

**Intellectual property rights (IPR)**—The impact of IPR on biotechnology research is often imbedded in discussions on public and private sector partnerships. There is a need to balance the fact that on one hand public sector institutions, due to limited resources, cannot fully avoid accessing private sector-held IP during the development of its own products and, on the other hand, the private sector has to avail of IPR protection to be able to protect its investments and commercial interests as well to be able to share their IP with other sectors without fear of exploitation. The development of the Golden Rice™ is a case in point. In total, 70 IPRs and technical property rights (TPRs) belonging to 32 different companies and universities were used in product development and for which 'freedom-to-operate' situations had to be applied for in order for NARES to begin using Golden Rice™ in further breeding and in de novo transformation activities using locally adapted varieties. Several modalities, however, are still open to the public sector to be able to access genes and technologies from the private sector. These include direct purchase of genes and technologies, licensing, the fact that patents have time limits, confidential agreements, and the purchase of genes for incorporation into local germplasm.

New types of IP agreements have also evolved, such as the donation of IP facilities and 'humanitarian' use type agreements as were done with Golden Rice™ with the threshold for humanitarian versus commercial use being a $10,000 income from the technology. The agri business giant Syngenta recently announced that it would donate new Golden Rice seeds and lines to the Golden Rice Humanitarian Board, including scientific results of the first field trials, as well as the technology, rights, and research results.

As the issue of IPR is likely to become increasingly important, the capacities of governments or the science sectors of many developing countries to understand, deploy, and negotiate regarding biotechnology need to be strengthened. Rice biotechnology practitioners in Asia need to be trained on the details of modern IPR systems and on negotiating with institutions and companies for the purpose of accessing IP, and applying for IP protection. Alternatively, research institutions could establish IP units, not only for negotiating with other institutions and sectors, but also for the IP registration of their own biotechnology processes and products.

**Regulatory requirements**—National biosafety committees in developing countries have made impressive progress in the drafting and implementation of biosafety regulations for the importation and testing of transgenic crops, with regulations for field tests already in place in rice-growing countries such as China, India, Thailand, and the Philippines. A looming issue, however, revolve around the compliance costs for regulatory approval that could prohibit many developing country institutions. In the various studies cited by Atanassov *et al.* annual compliance costs, including initial greenhouse and field screening, field testing for environmental impact, and food safety, but excluding technology development costs, ranged from US$140,000 for a virus-resistant papaya in Brazil to US$830,000 for a virus resistant potato in South Africa. For rice, an annual regulatory compliance cost of US$680,000 was estimated for a virus resistant variety in Costa Rica covering tests on molecular characterization and epidemiology, transgenic field trials, biosafety, IPR, food safety deployment, and gene flow. Given reduced NARES budgets, this could pose a major hurdle in the commercialization of rice biotechnology products from the public sector. It is hoped, however, that as knowledge and experience is gained by regulatory agencies, approval costs may decrease, both by reducing the number of required tests, and by shortening the length of experimentation. The latter would also avoid the risk of biotechnology products becoming irrelevant to farmers' needs due to approval delays. In this regard, continuous training of personnel from regulatory bodies of developing countries on new biotechnology developments and approaches is necessary for them to make educated recommendations, as is envisioned in the (CBD 2000). A well functioning regulatory system can hasten public acceptance of biotechnology products by instilling public confidence that the risk assessments conducted are carefully done, science-based, and, therefore, safe. While international harmonization of standards may be required, there is also need for contract-sensitivity that is appropriate to the place in which a technology will be applied.

**Policy support**—Policy initiatives are necessary to accelerate investments by technology holders and adoption by the farming communities in Asia. The Indian government, for example, is formulating new policies to boost investments and research in the local biotechnology sector. These include promoting the speedy approval of GM crops, and funding and infrastructure support for public-private partnership programmes plant biotechnology, among other areas. The new policies will also provide the framework for research and business institutions, and illustrate the trade and investment guidelines for the newly emerging biotech sector. A group of experts will also be set up to suggest models for public-private partnerships in the biotech sector. The biotechnology department will invest the creation of innovation centers with the existing academic and research institutions. To date, Delhi University has been identified to be the first center to receive the funding. Other cities such as Hyderabad, Pune, Chennai, Ahmedabad, and Lucknow will also be the focus of the development. Similarly, in Taiwan, Premier Yu Shyi-kun pledged his commitment to establish a Biotech Industry Strategy Consulting Committee, which would consolidate and integrate the country's biotech-based promotion organizations and research institutes.

Aside from policy initiatives, transparent protocols must also be established and international guidelines followed. There is also the need to establish a world-class intellectual property rights regime for biotechnology inventions and the protection of plant varieties. In addition, concurrent and transparent trials for biosafety for both new and released events are necessary and a determined drive against illegal genetically modified seeds trade is important.

**Biosafety and food safety**—Biotechnology is a very powerful tool that can be used in the developing world to grow more rice in an environmentally friendly manner. It can improve food production by making farming more efficient. The benefits that biotechnology confers upon the environment include the reduction in the use of agrochemicals and the preservation of presently uncultivated and marginal lands and their concomitant bio-diversity due to increases in productivity in the more favorable environments. To sustain the rice agriculture resource base and avoid environmental disturbance, it is important to match new genes and biotechnology-derived varieties to the conditions of the target environments. As commercialized GM crops by 2003 in the developing world was largely limited to insect protected cotton in Argentina, China, India, Mexico, and South Africa, there is still limited experience on food safety assessments to draw from for a food crop such as rice. Among developing countries, only three have approved a single transgenic event in a food crop (soybean in Brazil, Czech Republic and Uruguay; and maize in the Philippines), two have approved two events (soybean and tomato in Mexico; soybean and maize in South Africa) and one (Korea) has approved three events (one in soybean and two in maize). Therefore, the sharing of experiences and knowledge from the food safety assessments done in these countries should be valuable for developing countries with rice biotechnology products in the pre-commercialization stages. As rice is a food crop, and per capita rice consumption could vary both in-country and among countries, from less than 100 kg/yr in countries like China and India to over a 200 kg/year in countries like Myanmar,careful food safety experimentation must be done in the case of nutrient-enhanced GM rice to remove potential health dangers related to over dosages, if any, or alternatively, effective GM rice deployment strategies need to be developed.

**Communication**—Communication plays a vital role in disseminating information. It is a powerful tool in influencing man's decision. Media plays a crucial role in framing public understanding of science. Appreciation of traditional knowledge is essential for science to be communicated successfully. The message will be understood and better appreciated if there is an understanding of local knowledge. It is important to understand cultural values, and respect traditional knowledge to successfully communicate science. Scientists, journalists and others involved in the communication of science should take into account cultural factors and local knowledge in their work. The adoption and use of biotechnology and genomics should be approached within the context of conventional practices. Rather than view agricultural biotechnology as the wave of the future, we should regard it as part of a mélange of the new and old, balanced appropriately to meet local needs. The point was emphasized that agricultural biotechnology would be more readily adopted, and food security would be more attainable as would environmental sustainability, if it were blended back into conventional practices in order that valuea dded benefits would accrue alongside maintenance of traditions.

There are good examples of public education campaigns on biotechnology being done in Asian NARES. In Thailand, the Biotechnology Alliance Association (BAA) has been set up to educate the public about biotechnology applications, including genetically modified crops. Science based information will be provided to the public and several fora will be organized to discuss the benefits and risks of GM crops so that stakeholders can make intelligent decisions relative to biotechnology and its use. The potential for biotechnology to result in economic, environmental, and social benefits in India is enormous and promotion of the country's first hand experience in this area, through knowledge sharing activities, could serve as a powerful example for other developing countries. In the Philippines, the Biotechnology Association of the Philippines along with the Departments of Agriculture and Science and Technology, have also been mounting information campaigns through public dialogues, media presentations and regular press releases.

Given all these issues and challenges, lessons can be learned from the saga of Golden Rice™, as to the merging of factors necessary for a rice biotechnology product to be developed and commercialized for impact. As detailed by Potrykus, the project was made possible because of the enabling factors such as: (i) environment supportive of independent research; (ii) strong institutional collaborative research partnerships; (iii) availability of the needed genes; (iv) support from donor institutions for strategic research for developing countries; and (v) highly motivated team of scientists willing to work on a pro-poor R&D agenda. He further noted that the Golden Rice™ experience should: (i) facilitate greater public acceptance of GMO technology; (ii) encourage research investments in projects without guarantees for success, (iii) motivate research to be more food security- and less industry-focused; (iv) encourage free licensing for enabling technologies if used for humanitarian purposes; and (v) motivate scientists to undertake projects relevant to the poor.

The application of biotechnology in rice will help many Asian NARES to produce the estimated 700 million tonnes of rice required to feed an additional 650 million rice consumers by 2025. At the same time, it can also help lower the cost of rice farming, and add nutritional value to rice, thus benefiting resource-poor farmers through higher incomes and improved nutrition and health. Moreover, biotechnology may also be harnessed to protect the environment and sustain the natural resource base. The technology, however, must continue to draw its relevance from its being able to address both the existing and projected problems of small rice farming communities and, at the same time, the food and health needs of more than half of the world's population that consumes rice.

CHAPTER-6

# Promising Areas and Ventures

It is worth mentioning from both the technological and commercial viewpoints the case of biotechnology companies that use transgenic farm animals as bioreactors to produce life-saving medicines. If manufacturing biopharmaceuticals in animals could be made more efficient, it would ease the flow of new therapeutic compounds which is being held back. More than a hundred protein-based drugs are currently in advanced phases of clinical trials, and many more are in development in the laboratory. So far, Genzyme Transgenics Corporation (GTC) Biotherapeutics in Framingham, Massachusetts, has successfully engineered goats which produce 14 varieties of therapeutic protein in their milk. Creating a flock of transgenic goats costs about $100 million; this is expensive, but represents only a third of the cost of building a protein-production facility. Furthermore, when a drug maker needs to double production, the solution is to breed more animals, instead of spending $300 million on a new factory.

It generally takes about 18 months to make a transgenic goat that produces a desired therapeutic protein in its milk. This period is about three years for a cow, but milk production is higher (20 litres a day compared with two litres a day for a goat).

Regarding the production cost of the purified therapeutic

protein, it can be decreased to $1-2 a gram, compared with $150 a gram when the protein is extracted from cultured mammalian-derived cells. Chickens have some advantages over goats and cows. They lay eggs which are sterile, and the albumen or egg white, is an ideal storage medium for fragile compounds. They are faster to mature and cheaper to breed than goats and cows; a chicken flock can multiply tenfold within a year. Despite these advantages, research on transgenic chickens is less advanced than that on cattle or goats. In July 2002, TranXenoGen, based in Shrewsbury, Massachusetts, announced that it had produced two antibodies (one human and one murine) in the albumens of transgenic chickens. The yields of these antibodies must be increased and the company needed another year or so to achieve such result. TranXenoGen also aimed at producing transgenic chickens laying eggs containing insulin and human serumalbumin.

The few corporations which were breeding transgenic farm animals with a view to producing medicines hoped to pocket huge profits from sales estimated at billions of dollars over the coming decade. But they might be outpaced by those companies which have chosen to carry out the same process in crops such as maize, alfalfa, potato, etc., at a lower cost.

**Plant-derived drugs using molecular biology and biotechnology**

There is a long history in the use of plant materials and extracts for medical purposes. A classic example of a plant-derived drug is salicylic acid, found in willow bark, which is the basis for the use of acetylsalicylic acid, or aspirin, for relief of pain, fever and inflammation, and more recently for the protection against heart stroke. Plants with relevant medicinal properties were identified (botany) and the corresponding active compounds were purified and tested for both efficiency and safety (biochemistry and medicine). Then the best approach to production was determined, whether chemical synthesis, extraction from natural sources (intact plants, or tissue or organ cultures), or combination of both. Thus aspirin – a simple molecule with no chirality – was produced via chemical synthesis on an industrial scale.

Another example, more complicated than aspirin, is taxol – an isoprenoid compound found in the bark of the Pacific yew tree (*Taxus brevifolia*). It is highly successful in ovarian and breast cancer treatments. It is a complicated molecule, the total synthesis of which was reported in 1994 and involved over 40 steps. It was not commercially viable.

On the other hand, yew bark contains only 0.01% taxol, so supply is problematic. Production from cell cultures has been disappointing, although proprietary plant-cell fermentation technology recently developed may help. The best solution so far has been semi-synthesis. A precursor is obtained from yew needles and modified chemically to give taxol; this is not an entirely satisfactory solution and taxol stands out as a good example of a plant-derived drug that is difficult to supply.

Another approach is the application of molecular biology and biotechnology tools. With better knowledge of the genes involved in the biosynthesis of plant-derived drugs, their production may be enhanced in plants, plant-cell cultures or alternative hosts. This metabolic engineering approach is an area of active research. For instance, Rodney Croteau and colleagues at Washington State University are tackling the taxol problem in this way. Part of Canada's National Research Council's Programme on crop genomics led by Wilf Keller at the Plant Biotechnology Institute (PBI, Saskatoon, Saskatchewan) involves molecular genetic approaches to two other classes of plant-derived drugs with similar supply problems – aryl tetralin lignans and tropane alkaloids.

Podophyllotoxin belongs to the class of aryl tetralin lignans. It is an anti-viral drug and it is chemically modified to give three related anti-cancer drugs – Etoposide, Etopophos and Teniposide. Etoposide, for instance, has extensive application in the treatment of small cell lung cancer, advanced testicular cancer and Karposi's sarcoma. Other podophyllotoxin-related compounds have shown promise as anti-HIV agents. With four contiguous chiral centres and a high degree of oxygenation, total chemical synthesis of these compounds is not commercially viable, and the current source of podophyllotoxin is a herbaceous plant called *Podophyllum hexandrum* – a relatively rare plant that grows in the Himalayas and its current rate of harvest from the wild exceeds its regeneration rate.

A search for the relevant genes from *P. hexandrum* is being carried out at PBI. There is already fair knowledge about intermediates and enzyme-coding genes for the early part of the biosynthetic pathway from common phenylpropanoid precursors. Less is known about the latter part of the pathway. The Canadian researchers start by making a library that represents the genes in the tissue which makes the compound of interest. A few thousand are selected at random and sequenced. The sequences, called expressed sequence tags (ESTs), are compared against a large sequence data-base to give tentative identifications. Using additional data such as knowledge of gene expression in different tissues, the researchers can identify candidate cDNA clones which may encode enzymes of interest. These candidates are then tested by heterologous expression – the genes are introduced into a host such as *E. coli* or yeast, to test for activity from the enzyme of interest. The enzyme assays involved require good biochemical and analytical expertise because different and often unusual substrates and products are required for each assay.

The enzyme assays may point to the cDNA candidates corresponding to genes involved in podophyllotoxin biosynthesis. Once the genes have been found, it should be possible to metabolically engineer an appropriate host organism; a plant host may be the most feasible, but micro-organisms which can be easily contained may be more attractive than plants. Although one might wonder about the difficulty of introducing a number of genes at once into a plant, this is becoming more common. For instance, three genes were introduced into rice for the production of beta-carotene ('golden' rice) and up to twelve have been successfully introduced into soybeans. If all of the genes of interest in a biosynthetic pathway are not found, individual genes may be put to use in a few ways. One is to open metabolic bottlenecks in plants or cultures; the other possibility is the metabolic engineering of intermediate compounds with potential drug precursors.

Plants produce a wide range of secondary metabolites for defence and survival in their ecosystems. These secondary metabolites, currently exceeding 100,000 identified substances, belong to three major chemical classes: terpenes (a group of lipids), phenolics (derived from carbohydrates) and alkaloids (derived from amino-acids). The terpenes include, for instance, the diterpene taxol from the Pacific yew, and the triterpene digitalin from foxglove used as an effective drug for congestive heart failure.

The well-known phytoalexin, resveratrol, an anti-oxidant agent, is an example of phenolic as are flavonoids and tanmins which are found in tea, fruits, and red wine and have many desirable health effects. Alkaloids, a major class of plant-derived secondary metabolites used medicinally, have potent pharmacological effects in animals due to their ability to rapidly penetrate cell membranes. Nicotine, a commercially important alkaloid, is the most physiologically addictive drug used by humans. Caffeine, an alkaloid from coffee, tea and chocolate, is a central nervous system stimulant and mild diuretic. Vincristine and vinblastine, alkaloids from periwinkle, are strong antineoplastics used to treat Hodgkin's disease and other lymphomas. The opium plant contains over 25 alkaloids with morphine being the most abundant and most potent painkiller.

As a consequence of their numerous applications, the world market for plant extracts and isolated secondary metabolites exceeds $10 billion annually. The pharmacological value of plant secondary metabolites is increasing due to constant discoveries of their potential roles in health care and as lead chemicals for new drug development. However, secondary metabolites, generally present at 1% to 3% of dry plant biomass, are synthesized in specialized cells at distinct development stages and have highly complex structures, making their extraction and purification difficult. The content of secondary metabolites in biomass has been increased by cell-culture techniques in an effort to facilitate large-scale production, although the economic efficiency remains questionable. Henceforth, the idea to apply molecular biology and biotechnology tools to better understand the biochemical pathways leading to these compounds, isolate the genes coding for the relevant enzymes and metabolically engineer the producing plants or alternative hosts in order to increase the synthesis and output of secondary metabolites. The opium poppy (*Papaver somniferum*) has the unique ability to synthesize morphine, codeine, and a variety of other benzylisoquinoline alkaloids including papaverine and sanguinarine, which are of pharmaceutical importance. The global market for licit opium, from which the alkaloids are extracted, is in excess of 160 tons annually. Opium poppy produces low amounts of codeine due to a demethylase activity that converts codeine into morphine. Approximately 95% of the morphine extracted from licit opium is chemically converted to codeine, a more versatile pharmaceutical. The large quantities of the morphine produced by the plant are the basis for the illicit cultivation of opium poppy in many regions of the world in order to synthesize 0,0- diacetylmorphine, or heroin. The illicit production of opium exceeds licit usage by almost tenfold.

Research aimed at metabolic engineering in opium poppy could lead to biological alternatives for reducing the production and trafficking of illicit drugs around the world. Moreover, improved knowledge of secondary metabolism in opium poppy could also create opportunities to introduce entire pathways into value-added crops. In the case of benzylisoquinoline alkaloid biosynthesis in opium poppy, considerable progress has been made during the 1990s. No fewer than eight genes encoding alkaloid biosynthetic enzymes have been cloned from opium poppy. However, the biosynthesis of morphine and related alkaloids, such as the antimicrobial agent sanguinarine, involves many more enzymes. Moreover, much remains to be learned bout the control of alkaloid biosynthetic pathways, which are under strict regulation in plants. Consequently, our ability to harness the vast potential of these important secondary pathways is still rather limited. For instance, the use of plant organ, tissue and cell cultures for the commercial production of pharmaceutical alkaloids has not become a commercial reality despite decades of empirical research. Plant-cell cultures of opium poppy can be induced to accumulate sanguinarine, but do not produce morphine. This inability of dedifferentiated cells to accumulate certain metabolites has been interpreted as evidence that the operation of many alkaloid pathways is tightly coupled to the development of specific tissues. Recent studies have shown that alkaloid pathways, in general, are regulated at multiple levels, including cell type-specific gene expression, induction by light elicitors, enzymatic controls, and other poorly-understood metabolic mechanisms.

Advances in genomics will provide a more rapid and efficient means to identify biosynthetic and regulatory genes involved in alkaloid pathways. Novel insights obtained using a combination of conventional and modern techniques, including biochemistry, molecular biology, cell biology and genetic engineering, as well as advanced chromatographic methods to rapidly and accurately isolate, characterize and assess secondary metabolites, highlight the importance of a multifaceted approach to studying the regulation of alkaloid biosynthesis in plants, such as opium poppy.

### Biopharming: Plant-based pharmaceuticals

Epicyte Pharmaceutical of San Diego is one of a host of biotechnology companies that are involved in the production of plant-based pharmaceuticals. Researchers have launched more than 300 trials of genetically-engineered crops to produce everything from fruit-based anti-hepatitis vaccines to drugs against HIV/AIDS in tobacco leaves. Open-air trials of pharmaceutical crops (the process is called biopharming which can be related to medical biotechnology or to agricultural – 'green' – biotechnology) have taken place in 14 US States, from Hawaï to Maryland. Clinical trials have begun for experimental crop-grown drugs to treat cystic fibrosis, non- Hodgkin's lymphoma and hepatitis B.

Many researchers in North America, Europe and a few other parts of the world, are working to develop plant-based production systems for human therapeutic proteins. Crohn's disease is a chronic inflammatory bowel disease with discontinuous lesions occurring throughout the intestine tract. Current treatments relying heavily on corticosteroids and immuno-solocilates have multiple side-effects and complications.

There is 'therefore' a need for alternative therapies. Indications that cytokines retain some of their biological activity following oral administration, coupled with improvements in patients following parenteral administration of the anti-inflammatory cytokine, interleukin-10 (IL-10), led researchers at the Southern Crop Protection and Food Research Center, Agriculture and Agri-food Canada, London, Ontario, to believe that oral IL-10 may be an effective means of delivery to the gut. In order to produce sufficient low-cost IL-10 for oral administration, these researchers have expressed the human gene for IL-10 in low-nicotine tobacco. They have evaluated the biological activity of transgenic plants and intended to use them in an animal model of Crohn's disease.

William Langridge, professor in the department of biochemistry and at the Center for Molecular Biology and Gene Therapy at the Loma Linda University School of Medicine, California, and his colleagues have used transgenic potatoes to synthesize human insulin and showed that diabetic mice fed with these potatoes were less affected by the disease whose symptoms regressed. Douglas Russell of Monsanto Co. obtained tobacco plants producing human growth hormone (somatotropin) in its biological active form. Also tomatoes producing an inhibitor of the enzyme converting angiotensin-1 (i.e. an anti-hypertensive compound) were produced. Other plant-produced substances include: enkephalins, alphainterferon, serumalbumin and two of the most expensive pharmaceuticals – glucocerebrosidase and the factors stimulating the colonies of granulocytes and macrophages (GCMSF).

Glucocerebrosidase is an enzyme whose function is markedly decreased in patients suffering from Gaucher disease, which causes mental retardation in children, the retraction of large bones and the increase in size of liver and spleen. Another protein produced commercially in transgenic plants is hirudin, an anti-coagulant protein extracted from leeches, used to treat thrombosis. It has been produced in colza and mustard seeds by SemBioSys Genetics, Inc. Another commercial production example is that of human alpha 1-anti-trypsin in rice seeds by the California-based company, Applied Phytologics. This protein is used in the treatment of cystic fibrosis and haemorrhagia. Applied Phytologics hoped to obtain the authorization for commercializing this product in 2004.

Lactoferrin is the second most abundant protein in human milk, present at a concentration of 1 gram per litre. It is a multifunctional protein, which protects against microbial infection caused by a broad spectrum of bacteria, and may have an important role in the regulation of iron uptake in the gastro-intestinal tract as well as in the regulation of systemic immune responses through cytokine release. Fogher *et al*. of the Botany and Genetics Institute of the Catholic University S. Cuore, Piacenza, and Plantechno Srl, Cremona, Italy, have designed and transferred a synthetic lactoferrin gene into rice, with the aim to engineer a functional food to be used directly or as a vegetable milk in human nutrition. The synthetic human gene was optimized for codon usage in plants: as seed-specific promotors, the researchers have used the regulatory elements of the soybean storage proteins 7S globulin and beta-conglycinin with their own specific leader sequences. The transgenic rice lines were controlled for seed-specific lactoferrin production, iron content of the seeds and glycosylation pattern of the protein. Using the two rice varieties Ariete and Rosa Marchetti, a 82-kilodalton protein recognized by the anti-human lactoferrin antibody was produced in the seeds and not in the leaves. The amount of the protein accumulated exceeded 1% of total seed proteins and the plant lactoferrin was glycosylated at the same level as in the human protein.

In some transgenic rice lines, the quantity of seed iron was three times the normal concentration in the non-transgenic variety. SemBioSys Genetics, Inc., of Calgary, Canada, is using a variety of genetic engineering techniques to express proteins in the seeds of safflower. One embodiment of this technology involves the covalent attachment of proteins to oil bodies, natural oil-storage organelles found in oilseeds. Taking advantage of the physical principle that oil is lighter than water, oil bodies can be easily separated from the majority of other seed components. This provides a cost-effective solution for bulk protein production and purification. The technology is amenable for oral and topical delivery of bioactive peptides and proteins. The company operates a manufacturing facility that can deliver oil body-based products and purified protein at scale. SemBioSys Genetics, Inc., has been granted a US patent covering the use of its technology for producing somatotropins including human, bovine or fish forms of the hormone. This was the first example of the commercial-level expression of this class of proteins (correctly-folded and active disulfide-bonded proteins) in seeds. The fish somatotropin, thus produced, was physiologically active as an oil-body-associated protein when fed to salmon and trout, demonstrating its potential for oral delivery of biologically active proteins.

**Plantibodies and vaccines**

Although antibodies were first expressed in plants by the mid-1980s, the first report was published in 1989. Since then a diverse group of 'plantibody' types and forms have been prepared. Originally, foreign antibody genes were introduced into plant cells by nonpathogenic strains of the natural plant pathogen *Agrobacterium tumefaciens* and regeneration in tissue culture resulted in the recovery of stable transgenic plants.

Although this initial work to generate multichain proteins required crossing of plants expressing each chain, further studies have shown that multiple chains can be introduced via a single biolistic transformation event, greatly reducing the time to final assembled plantibody. James W. Larrick and colleagues of Palo Alto Institute of Molecular Medicine and of Planet Biotechnology, Inc., California, have transformed tobacco plants so that they produce antibodies against one of the surface proteins of *Streptococcus mutans* (agent of tooth decay); these plantibodies could successfully prevent the recolonization of the teeth and gums by this major oral pathogen in patients treated with the antibodies for at least four months.

At Thomas Jefferson University, the work carried out by Hillary Koprowski has shown that persons fed with transgenic lettuce containing an antigen of the hepatitis B virus had a good immune response against that virus. In 1998, Kevin Whaley and his colleagues, using soya bean plants transformed to produce antibodies against herpes virus simplex type 2 (HSV-2), succeeded in preventing vaginal infection by this virus in mice that had eaten the plants. This plantibodies were 100 to 1,000 times more efficient than other products used previously in this kind of experiment.

William Langridge is leading a team that genetically engineer potatoes to provide vaccination against cholera. The disease is contracted by more than five million people annually and causes more than 200,000 deaths worldwide. The researchers succeeded in transforming potatoes to produce the beta-subunit of cholera toxin (CTB). These potatoes were then used in two experiments. In the first, the team fed uncooked transformed potatoes to mice to develop the anti-CTB immunity. These mice were found to contain cholera antibodies in their blood and faeces. Also, those fed the biggest amount of potato were shown to be more resistant to the cholera toxin. When the effects of the vaccine were off, the mice were given a booster to maintain their immunity. In the second experiment, the researchers boiled the potatoes. An analysis of the boiled tissues showed that 50% of the CTB remained.

It is already known that CTB gives greater immunity against cholera to humans than it does to mice. However, existing vaccines do not adequately protect against this disease. The difficulty lies in the method of immunization. Vaccines are generally injected and stimulate an immune response in the bloodstream. But injections do not produce antibodies on mucosal surfaces, e.g. the walls of the gut. Vaccinations with plantderived vaccines (instead of calling them 'edible' vaccines) offer an alternative strategy to tackling this type of infection. Oral administration of transgenic-derived products will deliver cholera vaccine directly to the gut and provoke an immune response precisely where it is wanted. Langridge and his colleagues have yet to take their technology out of the laboratory.

Researchers at the Baylor College of Medicine in Houston have carried out trials on transgenic potatoes designed to protect against Norwalk virus, a major cause of water and food-borne diarrhoea in developing countries.

Vaccine-containing potatoes developed at the Boyce Thompson Institute for Plant Research (BTI), an affiliate to Cornell University, by Charles Arntzen and Hugh Mason, were also used in a human clinical trial. In the latter, led by Carol Tackett at the University of Maryland School of Medicine, Baltimore, volunteers were each served three helpings of raw potato to test the use of this kind of vaccine against traveller's diarrhoea. This common condition is transmitted by food or water contaminated by enterotoxigenic *Escherichia coli* (ETEC). The results of the trial, presented in the journal *Nature Medicine*, showed that individuals who had eaten the transgenic potato developed antibody protection against the diarrhoea. "Since infectious diseases cause a loss of more than 15 million kids annually, much of which could be prevented by good vaccines, it is our obligation to explore all new approaches to inexpensive and effective disease prevention", stated Charles Arntzen.

Charles Arntzen also has been working for nearly five years to create transgenic tomatoes containing a gene from a strain of *Escherichia coli* that can protect against diarrhoeal diseases. The US researcher focused on diarrhoea, because these diseases kill at least 2 million people in the world annually, most of them children. And he chose tomatoes because greenhouse-grown tomatoes cannot easily pass their altered genes to other crops and because tomato-processing equipment is relatively cheap. It would be easier to eat whole tomatoes, but that would be a disaster, according to Charles Arntzen.

Individual tomatoes come in different sizes with varying concentrations of the new protein, while uniformity of dosage is key to an effective vaccine. In 2002, Arntzen has tested juice derived from transgenic tomatoes on animals, with human trials to follow.

In relation to the research carried out at the Boyce Thompson Institute for Plant Research on plant-derived vaccines, studies have been made on storage protein promoting elements in banana, with a view to manipulating them for a better delivery of the vaccine. Clendennen *et al.* studied a banana protein (abbreviated P31) that appears to have evolved from an enzyme. P31 is abundant in the pulp of green bananas and it may play a role as a storage protein. P31 belongs to a class of proteins called the class III chitinases, several of which, although they are present during fruit ripening in avocado, cherry and tomato, have been thought to protect plants from disease or wounding; in certain plants, they are known to protect against fungi. In the banana, however, P31 chitinase decreases in abundance as ripening progresses. In general:

- storage proteins are very abundant (in unripe banana pulp, P31 accounts for approximately 20% to 30% of total soluble pulp protein);
- storage proteins are broken down during a subsequent developmental stage (P31 is broken down during banana ripening);
- storage proteins are generally localized in storage vacuoles within the cell (P31 is localized there);
- storage proteins contain a great proportion of particular amino-acid residues (P31 contains 22% of such residues, approximately the same as in soya beans and poplarstorage proteins – 21% to 25%);
- storage proteins typically lack any other metabolic or structural role in the organism; some retain a little enzymatic activity, but this could not be the case for the banana protein, which has only three of the five amino-acids required.

Clendennen *et al.* suspected that P31 served as a storage protein in banana and were carrying out experiments to determine whether a P31-promoting element introduced into tomato plants by genetic engineering would prove to be related to fruit ripening in the tomato, supporting the view of the role of P31. Storage-protein promoting elements might be genetically engineered to assist in vaccine delivery.

Some four dozen laboratories around the world are working on their own versions of plant-derived vaccines, using tomatoes, bananas and potatoes. The advantages of such vaccines, particularly in developing countries, are their cheap cost and their oral consumption instead of using needles (thus avoiding contamination due to the lack of rigorous asepsy). Attention is focused on tomatoes and bananas, which look likely to be more suitable for subtropical and tropical regions. Axis Genetics, based in Babraham, near Cambridge, Massachussets, is working in partnership with the Boyce Thompson Institute for Plant Research. The aim is to market vaccines against traveller's diarrhoea, Norwalk virus and hepatitis B in preserved foodstuffs or tablets.

With regard to regulatory aspects of plant-derived vaccines, an important consideration is that these vaccines must never occur as normal constituents of food, and should be produced under regulatory conditions to prevent contamination of food supplies and to maintain genetic containment. Furthermore, the vaccines will not be delivered in fresh form, but will be processed to yield uniform, stable batches with well-defined antigen content (henceforth the preference for food tablets containing the appropriate doses of antigen, rather than the raw genetically-modified vegetable or fruit). The vaccines will also be delivered by health-care professionals. Use of plants that are infertile and clonally propagated could facilitate management of quality control and production. Creation of male-sterile lines will enable more rigorous containment, because maternal inheritance of the chloroplast genome prevents pollen-mediated gene flow.

Charles Arntzen, now at Arizona State University, foresees rich markets for plant derived vaccines to protect fish and poultry against diseases currently being treated – and in many cases overtreated – with conventional antibiotics. For instance, clinical trials carried out by the biotechnology company, ProdiGene, Inc., have shown that the feeding of pigs with transformed maize containing a vaccine against transmissible gastroenteritis (TGEV) could protect them effectively against this disease.

Rinderpest is a highly contagious disease of cattle, buffaloes, sheep, goats and wild 92 ruminants, with a great mortality rate. Rinderpest virus has only one antigenic type (serotype) and an attenuated, live vaccine with high immunogenicity is available.

Rinderpest has been eradicated in developed countries, but is still prevalent in parts of Africa, the Middle East and South Asia, where eradication campaigns are underway.

The major drawback of the currently-used vaccine against rinderpest is its heat lability. In hot countries the vaccine delivery is constrained by high costs, lack of maintenance of cold chain to keep the potency of the vaccine. Although recombinant vaccines vaccinia/capripox recombinant or bacculo recombinants – have been produced, they have neither been tested in the field nor their usefulness in providing long-term immunity has been experimentally proven. Khandelwal *et al.* of the Indian Institute of Science Department of Microbiology and Cell Biology, Bangalore, have undertaken to develop transgenic tobacco plants (model system) and transgenic groundnut (*Arachis hypogea*) plants expressing the haemagglutinin (H) protein of rinderpest virus, as a source of vaccines delivered through the feedstuffs in order to immunize domestic ruminants as well as susceptible wildlife. The expression of H protein was demonstrated in the transgenic plants, and about 250 transgenic groundnut lines had been obtained by the Indian researchers.

**Preferred crop species**

So far, more than two-thirds of plant-based medicines were being tested in maize – a crop whose genetics is well understood. At Epicyte Pharmaceutical's laboratory tiny tobacco leaves, transformed with herpes-antibody genes, were grown in incubators. The Sacramento-based biotechnology company Ventria Bioscience broke new ground by planting 130 acres with new varieties of transgenic rice that will produce lactoferrin and lysozyme, to be marketed for use in oral rehydration products to treat severe diarrhoea. The company stated that this acreage could generate sufficient lactoferrin to treat at least 650,000 sick children and enough lysozyme for 6.5 million patients. It hoped to expand production to 1,000 acres within a few years.

The company did not disclose the site earmarked for the new crops because it was worried that protesters could destroy them. However, its plans have caused alarm in California's rice-growing community. Organic farmers, in particular, feared that transgenic rice could contaminate their crops. On 29 January 2004, the arguments were thrashed out before a meeting of the California Rice Commission, which was drawing up a protocol of conditions under which the transgenic rice varieties could be grown. In particular, the Commission was focusing on working out precautionary measures, e.g. the distance transgenic rice must be from conventional crops, to try to minimize the risks.

In the case of water lentils, the company LemnaGene LLC is specializing in genetically transforming plants of the Lemnaceae family through an agreement concluded with Bayer CropScience.

Based in Oregon, LemnaGene collaborates with the Weizmann Institute of Science and the Yeda Research and Development company in Israel. Transgenic plants will be used to manufacture functional foodstuffs, and new molecules for industrial, pharmaceutical and cosmetic uses. The advantages of *Lemna* spp. are their high productivity, a good knowledge of their genetics and transformation process, the possibility of growing them in hydroponic cultures in greenhouses (and not in the open air to avoid the escape of transgenes into the natural or agro-ecosystems). As it is not a food or industrial crop, the transformed water lentil cannot 'contaminate' conventional crops and may be preferred to maize for the production of drugs or other materials.

### Comparative economic advantages

Biopharming is mostly driven by a cost advantage. Building sophisticated factories to produce biopharmaceuticals can take as long as seven years and costs up to $600 million per facility. It is predicted that medicinal products could be synthesized in plants at less than one-tenth of the cost of conventionally-manufactured drugs and vaccines. These costs are, for instance, 10-to-50-fold less than protein produced at high concentration in *Escherichia coli* (i.e. 20% of total protein). Depending upon the use of the protein and the requirements for purification for *in vivo* pharmaceutical use, purification costs will obviously augment final product costs; however, at the hundred kilogram to metric ton level plant-produced proteins will provide obvious savings.

Consequently, demand for these low-cost products would grow rapidly and would far outstrip the capacity of conventional systems to produce them. Where pharmaceutical factories produce Interleukin-10 in kilogram quantities, farmers would be able to produce it literally by the ton.

By the end of the current decade, biopharmaceuticals were projected to grow into a $20- billion industry. But how many of the new drugs will be manufactured in plants remains uncertain. This technology could bring down the cost of treating a number of diseases in a significant way, so that the drawbacks compared with the benefits will be very small.

**Pharmaceuticals and nutraceuticals from marine organisms**

The large pharmaceutical groups seem to abandon their search for new drugs derived from natural substances, because over the 1990 decade only one out of 10,000 to 20,000 compounds extracted from terrestrial micro-organisms, plants or animals could become an effective medicine. This may not be the case for marine organisms as relentlessly stated by José María Fernández Sousa-Faro who founded the first Spanish pharmaceutical company PharmaMar – and the only existing one up to 2004 –, a subsidiary of the Zeltia group, which is devoted to seeking new drugs from marine organisms. It is among the half dozen companies across the world which carry out this kind of research, and to that end it has close ties with the fisheries group, Pescanova, which owns fishing vessels and screening centres in all oceans.

Located near Madrid, PharmaMar has been working for 17 years on research and development of anti-tumour products derived from marine organisms. Its staff has increased from 70 to 300 during the three-year period 2000-2002. Its new facilities, inaugurated in March 2003, required a €22-million investment and included a pilot plant for the production of drugs.

In 1996, PharmaMar's president announced that a compound derived from a marine Tunician, ET-743 (ecteinascidin-743), was to be tested in clinical trials for its potent anti-tumour activity. Two years later, at the Congress of the European Organization for Research and Treatment of Cancer (EORTC) in Amsterdam, J.M. Fernández Sousa- Faro reported the initiation of phase-2 clinical trials of the same compound, renamed Yondelis. This became the star product of the company.

At the same time, PharmaMar went public and was able to gather €230 million in the stock market, of which €120 million have been devoted to developing the first Spanish anti-tumour compound. This was very unusual in Spain, because a company has never funded its research and development (R&D) with the money collected in the stock market. NeuroPharma or PharmaGen – the other subsidiaries of PharmaMar – are willing to follow suit, which is a very good notice for the Spanish R&D system that lacks really innovative firms. By contrast, in the USA it is common to use funds gathered from the stock market to support innovative research, while in Spain most corporations have to rely on the public assistance provided by the Centre for Technological and Industrial Development (CDTI), the Interministerial Commission on Science and Technology (CICYT), the Programme for the Promotion of Technological Research (PROFIT) of the Ministry of Science and Technology, as well as the European Regional Development Funds (FEDER).

Since its creation and up to 2004, some €300 million have been invested in PharmaMar, out of which €120 million have been allocated to research on Yondelis. During that period, four products have been tested in clinical trials and another 14 were in a preclinical stage. The €230 million collected in 2000 in the stock market have fueled the research and by the end of 2003, €130 million were left for funding it for another two years. PharmaMar's president was hoping that during this period a breakthrough would occur regarding Yondelis.

Unfortunately the European Agency for the Evaluation of Medicinal Products (EMEA) has given a negative judgment in the first evaluation phase of Yondelis, despite the fact that PharmaMar presented at the EORTC congress in Amsterdam all the details concerning the anti-tumour activity of Yondelis, and after the failure of the two drugs generally used against sarcomas of soft tissues, doxorubicine and phosphamid. While over the 20-year period up to 2003, almost 50 products have been tested against this kind of sarcoma without success, Yondelis was the first compound that showed a specific activity. Soft-tissue sarcoma is a rare disease, representing only 1% of all forms of cancer, and this implies that it is not easy to find patients on whom any potential drug could be tested.

At the initiation of phase-2 clinical trials, three patients died, because 50% of the product is metabolized in the liver and the rest is eliminated through the bile. The patients with obstructed bile ducts did not eliminate the product, so that the dosis was higher than the tolerable one. It is easy to know whether the patient has a non-obstructed bile duct (through a test of alkaline phosphatase and bilirubine); if this is not the case, instead of 1,500 μg/m² of Yondelis, 1,200 μg/m² are administered. With this change in the treatment protocol, there were no problems. In addition, there was no cumulative toxicity, and some patients have received more than 30 cycles, while with the other available drugs, one cannot go beyond six cycles. Despite the setback at the EMEA level, PharmaMar's president remains optimistic about the future of the company's star product. In addition to its activity against soft tissue sarcoma, Yondelis was found to also have a specific activity against ovary cancer, particularly among women whose cancer is resistant to the usual treatments with taxanes and cisplatinum. Yondelis could be an alternative therapy and new studies were being designed. PharmaMar is being consolidated despite this blow and is really innovative in its development of drugs derived from marine organisms. One is dealing with new compounds having also a novel anti-tumour mechanism of action, which may pave the way to new modes of action against tumoural cells. The near future will tell if PharmaMar can succeed in what is generally considered a risky venture. New Zealand has an extensive coastline, and a diverse and unique algal flora, including 800 known species of seaweeds. This represents an ample source for exploration. Industrial Research Limited (IRL), a New Zealand Government Crown Research Institute, and New Zealand Pharmaceuticals (NZP) have formed a strategic alliance with the Hobart-based company, Marinova Pty Limited, with a view to commercializing marine organisms-derived compounds for pharmaceutical and nutraceutical applications. The IRL was set up in 1992, following the restructuration of the Directorate of Scientific and Industrial Research (DSIR). This restructuration resulted in the formation of several Crown Research Institutes. The IRL is owned by the New Zealand Government and overseen by a board of directors, which includes representatives from several high-profile New Zealand based-companies. The IRL undertakes contract research-and-development projects in technology areas, including advanced materials, biochemical engineering, complex measurements and analysis, alongside commercialization activities. It also has the capability of specialty manufacturing within IRL Biopharm and IRL Glycosyn. The carbohydrate chemistry team at IRL is recognized as a world leader in the development of high-value carbohydrate compounds.

With new facilities, backed by 20 years in seaweed research, the development of novel compounds has now become commercially viable, through a strategic alliance between New Zealand and Tasmania – a trans-Tasman link. Thus, New

Zealand Pharmaceuticals (NZP) now manufactures Marinova's bioactive compounds and is a shareholder in Marinova. This close relationship brings significant expertise in process development under good manufacturing practice (GMP) standards. The focus of Marinova, IRL and NZP's seaweed programme is on high-value products. Initial commercialized products are for the dietary supplements market with prospects to launch botanical drugs in the short to medium-term under a new US Federal Drug Administration (FDA) category.

The research focus to date has been the polysaccharides (derived form hydrocolloids) in seaweeds (often sulphated galactans), which have a long history of being used commercially. Agars, carrageenans and alginates are all useful compounds that cannot be made artificially, so they have to be extracted from the cell walls of seaweeds. The New Zealand government has recognized a potential for creating a new high-value industry from seaweeds. This is reflected in a commitment of NZ$7 million dollars over six years to the IRL, leveraging the substantial investments of Marinova, NZP, other research providers and industry stakeholders.

Late in 2002, following five years of development, Marinova, IRL and NZP commercialized a novel compound – a polysaccharide from the seaweed *Undaria pinnatifida*, an introduced species harvested in Tasmania. Native to Japan and parts of Asia, where it is commonly known as wakame, *Undaria* has also been introduced to New Zealand, South America and parts of Europe. Marinova, through its subsidiary Marine Resources Pty Limited, developed a commercial harvesting and processing model, built around containment of this invasive species of seaweed. The Tasmanian Department of Primary Industries, Water and Environment, together with the Department of Economic Development, have been fostering this new industry.

Commercial harvesting of *Undaria* began in late 2002 and over 200 tons of biomass was harvested by divers during the first commercial harvest season, compared with previous pilot quantities of up to 60 tons. 'Viracle with GFS™' was Marinova's first product, entering the US market as an anti-herpes therapy. Galacto Fucan Sulphate (GFS™) was proven a potent anti-viral *in vitro* and in human pilot trials. Marinova has made considerable investment in clinical research, trials and regulatory compliance with a number of markets including Canada, United Kingdom, Australia and New Zealand.

The company had plans to be present on these markets in 2004. The focus of the Tasmanian and New Zealand researchers includes an evaluation of marine biological diversity, marine pest strategies, marine farming of high-value seaweeds and biosafety. New Zealand Pharmaceuticals Limited (NZP) has been purifying biochemicals from natural raw materials since the early 1970s. It supplies a range of biochemicals to the pharmaceutical, cosmetic, health-food and biotechnology industries with over of 95% of its products being exported. NZP has changed its focus from the production of biochemicals from meat industry by-products, to those derived from plant materials. NZP has a long history of manufacturing polysaccharides, including heparin and the related glycosaminoglycan, chondroitin sulphate. Experience in the extraction and purification of these animal-derived polysaccharides led the company to develop new technologies for extracting marine polysaccharides in conjunction with Marinova. NZP was able to quickly convert the initial information provided by Marinova and IRL into a commercial extraction process for *Undaria* harvested in Tasmania.

New Zealand's commitment to creating a high-value industry from seaweeds highlights the importance of biotechnology to this country, as shown in the report 'Growing the biotechnology sector in New Zealand – A framework for action', presented to the government and released to the public on 6 May 2003 by the Biotechnology Task Force, set up under the new Zealand Government's Growth and Innovation Framework in 2002 and comprising of members from business, universities and Crown Research Institutes.

The Task Force's report, produced in cooperation with the biotechnology sector, Industry New Zealand and the Ministry of Research, Science and Technology, outlined actions for commercializing biotechnology innovations, building the critical mass of the sector and removing barriers to growth. Key targets in the report included:

- creating an industry with NZ$10-billion market capitalization;
- tripling the size of the New Zealand biotechnology community to over 1,000 organizations from 350;
- increasing total cluster employment to over 18,000 from around 3,900;
- raising five-fold the number of core biotechnology companies to over 200 from 40. A global focus by biotechnology business and action by government to increase international awareness of New Zealand's strengths was also stressed by the report. A joint initiative between the New Zealand Government and bio-industry, the website 'Biospherenz.com' was designed to provide information aimed at international investors and researchers.

An independent study by the Channel Group identified nine key biotechnology sectors that offered great potential for growth in New Zealand: biopharmaceuticals, biomanufacturing, agricultural biotechnology, transgenic animals, bioactives, industrial and environmental biotechnology, nutraceuticals, and clinical research and trials.

**Cosmeceuticals**

A new market niche known as cosmeceuticals – products that are marketed as cosmetics but contain biologically-active ingredients – is proving to be lucrative. Cosmeceuticals indeed are one of the fastest-growing segments of skin-care business: in 2003, total skin-care sales in department stores grew 6%, while sales of cosmeceuticals and clinical brands jumped 77%; in 2002, total skin-care sales were up to 4%, while sales of cosmeceuticals and clinical brands rose 62%. Cosmeceuticals are so appealing because they are more affordable than Botox or Restalyne, which are injected into facial muscles to erase wrinkles, and yet enjoy medical credentials. According to Aurelian Lis, cofounder of Prescribed Solutions – a New York-based cosmeceuticals company – 'cosmeceuticals offer some of the benefits of pharmaceuticals but are still inherently cosmetics'.

Big cosmetic groups have enlisted the help of dermatologists. Lancome, a division of L'Oreal – the world's biggest cosmetics group with 18.7% of the €70-billion global cosmetics market and about €14-billion sales in 2003 – hired a specialist in dermatologic laser surgery as a consultant in September 2003, while Prescriptives – a unit of Estee Lauder – recruited a dermatologist in October 2002. In 2003, Estee Lauder bought the Rodan & Fields skin-care line developed by two dermatologists. Virginia Lee, US research analyst at Euromonitor International, a market research group, stated cosmeceutical manufacturers are positioning their products as an option before taking more drastic steps such as plastic surgery, Botox injections or chemical peels.

As cosmeceuticals have become more mainstream, some products have appeared in stores, in addition to being located in physicians' offices. NV Perricone MD Cosmeceuticals are found in upscale retailers. Sephora, owned by the world's largest luxury-goods group, Louis Vuitton-Moet-Hennessy, is at the forefront of the cosmeceutical trend; its US stores sell several cosmeceutical brands, including NV Perricone, Dr Murad and Dr Brandt. Sales at NV Perricone more than doubled in 2002 to \$42.4 million from \$11.9 million the year before.Overall cosmeceutical sales are rising. A report by the Freedonia Group, a market research firm, stated that the \$3.4-billion cosmeceutical industry was poised to grow 8.5% to \$5.1 billion by 2007. While it is a fraction of the \$33-billion US cosmetics and toiletries market, cosmeceutical sales are growing much faster than the overall market.

This trend has gained momentum due to ageing baby boomers who want to look young and further to Botox approval in 2002 by the US Food and Drug Administration. More than 1.6 million people in the USA removed wrinkles with injectable treatments in 2002, according to the American Society of Plastic Surgeons, while more than 4 million Americans had non-surgical cosmetic treatment. Allergan, which makes Botox, reported that fourth-quarter sales of Botox rose more than 20% to \$158 million. For 2004, the company expected Botox sales to grow to \$660 million-\$700 million – up as much as 24% from \$563.9 million in 2003.

CHAPTER-7

# Applications of Biotechnology for Functional Foods

Foods qualify as 'functional foods' because they contain non-essential substances with potential health benefits. Examples of the diverse foods and their bioactive substances that are considered 'functional foods' are: psyllium seeds (soluble fiber), soya foods (isoflavones), cranberry juice (proanthocyanidins), purple grape juice (resveratrol), tomatoes (lycopene), and green tea (catechins). The broad classification of functional foods carries some irony, as John Milner, Chief of the Nutrition Science Research Group at the National Cancer Institute noted, "It is unlikely that a non-functional food exists.

" Bioactive components of functional foods may be increased or added to traditional foods through genetic engineering techniques. An example would be the high lycopene tomato, a genetically modified tomato with delayed ripening characteristics that is high in lycopene, which has potent antioxidant capabilities. This report focuses on biotechnology applications in functional and improved foods, using the National Academy of Sciences definition as a guideline.

## Applications of biotechnology in food crops

In 1990, the U.S.Food and Drug Administration (FDA)

approved the first genetically engineered food ingredient for human consumption, the enzyme chymosin, used in cheesemaking. It is estimated that today 70% or more of cheese made in the U.S. uses genetically engineered chymosin. The first genetically engineered food, the FlavrSavr™ tomato, was approved for human consumption in the U.S. in 1994.

## Transgenic acreage expands steadily

Seven million farmers in 18 countries now grow genetically engineered crops. Leading countries are the U.S., Argentina, Canada, Brazil, China, and South Africa. Cultivation of genetically engineered crops globally has expanded more than 10% per year for the past seven years, according to the International Service for the Acquisition of Agri-biotech Applications ISAAA.. Such an expansion rate amounts to a 40-fold increase in the global area of transgenic crops from 1996 to 2003. Thus, in spite of continuing controversy, the technology continues to be adopted by farmers worldwide. ISAAA highlighted its key findings this way:

In 2003, GM crops were grown in 18 countries with a combined population of 3.4 billion, living on six continents in the North and the South: Asia, Africa and Latin America, and North America, Europe and Oceania.... the absolute growth in GM crop area between 2002 and 2003 was almost the same in developing countries (4.4 million hectares) and industrial countries (4.6 million hectares) ... the three most populous countries in Asia—China, India, and Indonesia, the three major economies of Latin America—Argentina, Brazil and Mexico, and the largest economy in Africa, South Africa, are all officially growing genetically engineered crops. The leading genetically engineered crops globally and in the U.S. are soya, maize (corn), cotton, and canola. In the U.S., transgenic virus-resistant papaya and squash are also cultivated.

## Agronomic traits prevail

Research in plant biotechnology has focused primarily on agronomic traits—characteristics that improve resistance to pests, reduce the need for pesticides, and increase the ability of the plant to survive adverse growing conditions such as drought, soil salinity, and cold. Biotechnology traits developed and commercialized to date have largely focused on pest control (primarily Bt crops) or herbicide resistance. Many plant pests have proven either difficult or uneconomical to control with chemical treatment, traditional breeding, or other agricultural technologies and in these instances in particular, biotechnology has proven to be an effective agronomic tool. Herbicide resistance allows farmers to control weeds with chemicals that would otherwise damage the crop itself. Varieties combining two different traits, such as herbicide tolerance and insect resistance, havc bccn introduccd in cotton and corn. The addition of new traits, such as resistance to rootworm in maize, and the combinations of traits with similar functions, such as two genes for resistance to lepidopteran pests in maize, are expected to increase. In its 2003 report, ISAAA suggested that five new Bt and novel gene products for insect resistance in maize could be introduced. While the improvement of agronomic characteristics in major crops has been highly successful, few products genetically engineered to meet the specific needs of either food processors or consumers have yet been commercialized. Recently, however, a renewed emphasis on developing agricultural biotechnology applications more relevant to consumers has accompanied continuing efforts to develop crops with improved agronomic traits. Although genetically engineered crops with enhanced health, nutrition, functional, and consumer benefits have lagged behind agronomic applications, research on many such products is in the advanced stages of development.

These applications could improve human and livestock nutrition and health, the nutritional quality of food animals for human consumption, and create ingredients with superior properties for food manufacturing and processing.

## Food applications for human health

### Quantity and quality of food oils

Food oils have both nutritional and functional qualities. From a nutritional perspective, fats and oils contribute more energy (calories) than any other nutrient category, about nine calories per gram. This compares with about four calories per gram from carbohydrates and protein. At the same time, specific fatty acids that comprise most of what we call 'fat' can affect a person's risk of developing certain chronic diseases such as heart disease. Research over the past several decades has shown that some categories of fatty acids, such as saturated fatty acids, increase the risk of heart disease and other chronic diseases when consumed in excess. Fatty acids also influence how foods behave during manufacturing and processing. For example, saturated fatty acids add stability, texture, and flavour to foods, so they are not simple to replace. To reduce the saturated fatty acid content of foods, plant breeders and food manufacturers increased their use of vegetable oils rich in polyunsaturated fatty acids and developed food oils low in saturated fatty acids. One example is canola oil with 6% to 7% total saturated fatty acids. To improve the stability of vegetable oils rich in polyunsaturated fatty acids, food manufacturers developed partially hydrogenated oils. The process of hydrogenation reduced the polyunsaturated fatty acid content and increased oil stability, but created *trans* fatty acids, which were subsequently associated with adverse health effects. As a result, hydrogenated fats, the main source of dietary *trans* fatty acids, are now being eliminated from foods. Food manufacturers are developing other ways to reduce undesirable saturated fat content while maintaining stability such as using short chain saturated fatty acids and monounsaturated fatty acids.

To date, one functional food oil created with the tools of biotechnology has been commercialized. Calgene's high lauric acid canola, Laurical™, containing 38% lauric acid, is used in confectionary products, chocolate, and non-food items such as shampoo. Conventional canola oil does not contain lauric acid. Laurical™ is a substitute for coconut and palm oils. FDA approved its use in foods in 1995. The following section describes research to date focused on developing crop varieties with other unique oil profiles.

## Strategic aims of altered fatty acid profile

Improving the healthfulness and functionality of food oils can be accomplished in several ways. Where traditional plant breeding reaches its limits, biotechnology may be used to:

- reduce saturated fatty acid content for 'heart-healthy' oils,
- increase saturated fatty acids for greater stability in processing and frying,
- increase oleic acid in food oils for food manufacturing,
- reduce alpha-linolenic acid for improved stability in food processing,
- introduce various omega-3 polyunsaturated fatty acids including long-chain forms and
- enhance the availability of novel fatty acids, e.g., gamma-linoleic acid.

## Achievements in altered fatty acid profile

*Reduced saturated fatty acid content*: Genetically modified soya beans have been developed that contain about 11% saturates compared with 14% in conventional soya beans (Table 1). In May 2003, scientists reported the development of transgenic mustard greens (*Brassica juncea*) containing 1% to 2% saturated fatty acids, a level significantly less than in the control plants. The transgenic plants also contained slightly higher amounts of oleic acid, a monounsaturated fatty acid, and higher levels of the polyunsaturates, linoleic and alpha-linolenic acids than the control plants. These results illustrate that alterations in one type of fatty acid may affect the levels of others, suggesting that combined strategies or genetic transformations may be necessary to achieve specific fatty acid profiles.

Palm oil low in saturated fatty acids is currently in development. This tropical oil contains about half saturated fatty acids (49.3%), primarily palmitic acid (16:0, 43.5%). However, with the recent success of biotechnology techniques in palm, transgenic palm oil enriched in oleic and stearic acids is under development. Because of the long life cycle of palm and the time required to regenerate the plants in tissue culture, genetically engineered palm is not anticipated for another two decades.

*Increased saturated fatty acid content*: Because saturated fatty acids confer certain functional properties to food fats and oils and are more stable to heat and processing than unsaturated fatty acids, their use in cooking and baking is essential. To avoid the use of animal fats and hydrogenated vegetable oils with trans fatty acids, genetic engineering techniques have been used in canola and soya to develop oils with more short chain saturated fatty acids—12 to 18 carbons long—mainly lauric (12:0), myristic (14:0), palmitic (16:0), and stearic (18:0) acids. For example, Calgene's high lauric acid canola, Laurical™, containing 37% lauric acid, was developed using the enzyme acyl-ACP thioesterase isolated from the California Bay Laurel (*Umbellularia californica*). Conventional canola oil contains no lauric acid, and only about 6% short chain saturated fatty acids. This was the first transgenic oilseed crop produced commercially. High laurate canola is used in confectionary products, chocolate, and non-food items such as shampoo as a substitute for coconut and palm oils. FDA approved its use in 1995.

Enrichment of canola with even shorter chain saturated fatty acids, those with eight and ten carbons, has also been accomplished. Using a palmitoyl-acyl carrier protein thioesterase gene from a Mexican shrub, *Cupea hookeriana*, Dehesh and colleagues developed lines of canola with as much as 75% caprylic (8:0) and capric acids (10:0). These fatty acids are absent in conventional canola oil. When consumed, these water-soluble fatty acids are mainly oxidized for energy.

Soya beans have been genetically modified to produce oil enriched in stearic acid (18:0), a saturated fatty acid that scientists believe does not raise serum cholesterol levels. The stearic acid-rich oil shown in Table 1 had 28% stearic and 20% oleic acids, with lower linoleic acid (18:2) than the conventional oil. Gene transfer technology also boosted the stearic acid content of canola. Researchers at Calgene, Inc., Davis, CA, cloned three thioesterase genes from mangosteen, a tropical tree that stores up to 56% of its seed oil as stearate. One of these genes led to the accumulation of up to 22% stearate in transgenic canola seed oil, an increase of more than 1,100% over conventional varieties.

*Increased oleic acid content*—The most recent approach to developing more healthful food oils is increased oleic acid content. High oleic acid oils are lower in saturated and polyunsaturated fatty acids compared with conventional oil. Oleic acid, the predominant monounsaturated fatty acid in seed oils, is abundant in olive (72%), avocado (65%), and canola (56%) oils, but not in others. Like saturated fatty acids, high oleic acid oils are useful in food processing and manufacturing for maintaining functionality and stability during baking and frying. Unlike saturated fatty acids, however, they do not raise blood cholesterol concentrations and are 'therefore' considered more healthful.

Biotechnology offers a means to increase the oleic acid content of vegetable oils, usually at the expense of polyunsaturated fatty acids, and sometimes, saturated fatty acids, depending on the particular transformations used. The concomitant reduction in polyunsaturated fatty acids has the added advantage of increasing the stability of the oil and ultimately the processed food. While traditional plant breeding allowed a modest increase in oleic acid, biotechnology has been necessary to achieve the high levels desired. For example, canola oil moderately high in oleic acid was developed using traditional plant breeding techniques. With the application of biotechnology, oleic acid content increased to 75%. Others have developed canola oil with more than 80% oleic acid.

More recently, Buhr and colleagues at the University of Nebraska used genetic engineering to increase oleic acid levels in soybeans by inhibiting the ability of the plant to convert oleic acid to polyunsaturated fatty acid. When the conversion enzyme was inhibited, the level of oleic acid increased from 18% in the wild-type seed to 57% in the transgenic seed. When two gene transformations were applied, oleic acid content increased to 85%, with saturated fatty acids reduced to 6%.

Using a different approach, scientists at DuPont used the technique of cosuppression to reduce the production of polyunsaturated fatty acids in soybeans. Cosuppression occurs when the presence of a gene silences or turns off the expression of a related gene. Like Buhr and colleagues, these scientists were able to turn off the production of the enzyme that converts oleic acid to polyunsaturated fatty acids. The result was greatly increased production of oleic acid and reduced production of polyunsaturated fatty acids. Examples of genetically modified high oleic acid oils compared with their conventional counterparts are shown in Table 1.

Gene silencing has also been used to produce high oleic and high stearic acid cottonseed oils. Cottonseed oil is high in palmitic acid, very high in linoleic acid, and free of alpha-linolenic acid. Conventional cottonseed oil has about 13% oleic acid. When gene silencing was used to transform cotton, the resulting oil had 78% oleic and only 4% linoleic acids, respectively, with palmitic acid reduced from 26% to 15%. Cotton was also genetically modified to produce high stearic oil having 40% stearic and 39% linoleic acids, with 15% palmitic acid. A combination was also developed to have 40% stearic, 7% oleic and only 6% linoleic and 14% palmitic acid. These examples illustrate the power and specificity of this technology to develop tailored seed oils.

*Reduced alpha-linolenic acid*—Several genetic transformations designed to increase oleic or stearic acid content do so at the expense of the polyunsaturated fatty acids alpha-linolenic and linoleic acids. These fatty acids have desirable nutritional characteristics, but their presence reduces the stability of oils for baking, processing, and frying and increases their susceptibility to oxidation or rancidity. Oils with appreciable amounts of alpha-linolenic acid such as canola and soya bean, with about 10% and 8% alpha-linolenic acid, respectively, have been genetically modified to reduce this fatty acid. Such oils would be desirable for the commercial uses mentioned. Pioneer Hi-Bred, a DuPont company, developed low alphalinolenic acid soya bean seeds through conventional breeding techniques with less than 3% alpha-linolenic acid in its oil. Marketed under the brand TREUS™ the company claims that the oil eliminates the need for hydrogenation in food processing. A similar product from Monsanto, Vistive™, offers a similar level of reduction in alpha-linolenic acid.

*Omega-3 fatty acids*—There is extensive interest in increasing Americans' consumption of omega-3 fatty acids, because they are associated with many health benefits, but are consumed only in small amounts. In 2002, the National Academy of Sciences' Institute of Medicine recognized that omega-3 fatty acids are essential in the diet and established an estimated adequate intake for them. The main food sources of the long-chain omega-3 fatty acids are fish, especially fatty species such as salmon, rainbow trout, mackerel, herring, and sardines. Some plants—mainly canola, soybean, and flax oils—provide the 18-carbon omega-3 fatty acid, alpha-linolenic acid. However, higher plants lack the enzymes to make 20- and 22-carbon polyunsaturated fatty acids needed by mammals. Humans can convert alpha-linolenic acid to the more biologically active longchain forms, but they do so very inefficiently. Thus, plant foods with alpha-linolenic acid may be insufficient to supply the need for long-chain omega-3 fatty acids, especially during pregnancy and lactation.

Western diets contain predominately omega-6 polyunsaturated fatty acids found in soya bean, corn, sunflower, canola, and cottonseed oils. It is now recognized that diets high in omega-6 fatty acids and low in omega-3 fatty acids may exacerbate several chronic diseases. Because of the many health benefits associated with the regular consumption of omega-3 fatty acids, several health organizations, including the American Heart Association and the 2005 Dietary Guidelines for Americans, have called for increased consumption of these substances. One limitation to boosting consumption is that they occur naturally mainly in fatty fish and some seeds. Ironically, reducing the level of alpha-linolenic acid in soy and canola oils used in food processing, may actually reduce consumption of this fatty acid, although product developers are working to combine high omega-3 and low alpha-linolenic traits in one product.

Although aquaculture has increased the availability of some fish and shellfish species, increasing worldwide demand has put severe pressure on wild aquatic resources and limited seafood availability. Thus, it would be desirable to increase the availability of these fatty acids or their precursors in a variety of other foods, especially plants. Such foods would also be useful for animal and fish feed.

One strategy to increase the availability of long-chain omega-3 fatty acids is to develop oilseed crops such as canola and soya bean that contain *stearidonic acid* (18:4n-3). This omega-3 fatty acid occurs naturally in only a few plants such as black currant seed oil and echium. Stearidonic acid is the first product formed when alpha-linolenic acid is converted to eicosapentaenoic acid (EPA), a desirable long-chain omega-3 fatty acid. Usually, this first step limits the amount of EPA produced, but increasing the level of stearidonic acid helps overcome this limitation. Then the body's enzymes convert stearidonic acid to 20-carbon polyunsaturated fatty acids.

Dr. Virginia Ursin and colleagues at Calgene studied the metabolism of stearidonic acid in people. Her studies showed that when either stearidonic acid or EPA was consumed the amount of EPA in red blood cells increased significantly. This finding meant that the stearidonic acid was converted to EPA and appeared in red cells just as readily as the preformed EPA. In contrast, when the study volunteers consumed alpha-linolenic acid, there was no change in their red cell EPA content. None of the fatty acids consumed had any effect on cell DHA levels, another long-chain omega-3 fatty acid associated with health benefits. Although the study used supplements, not stearidonic acid from transgenic plants, the findings suggest that plants with stearidonic acid would have potential to provide EPA.

Toward this end, scientists at Calgene, have successfully transformed canola so that it makes stearidonic acid. This genetic engineering feat required two genes from the fungus *Mortierella alpina* and one from canola for the three enzymes needed to produce sufficient stearidonic acid. The engineered plants accumulated up to 23% stearidonic acid in the seed oil with a reduction in oleic acid content from about 60% to about 22%. By breeding the transgenic lines with various lines of canola the investigators were able to develop a line of canola containing more than 55% of alpha-linolenic acid and stearidonic acid. Total omega-6 fatty acids remained about 22%, a level similar to conventional canola. Calgene scientists have also developed soya bean that contains stearidonic acid.

The implications of Calgene's work with stearidonic acid are substantial. This is the first demonstration of the incorporation into edible plants of a biologically potent source of long-chain omega-3 fatty acids. This work marks an important advance in the development of plant-based sources of long-chain omega-3 fatty acids that could be consumed directly or incorporated into food products. However, because stearidonic acid contains four double bonds, it is vulnerable to oxidation and would require antioxidant protection. One can imagine that transgenic canola and soybean could be developed using additional traits to boost antioxidant protection, possibly from vitamin E.

In May 2004, a landmark paper announced the production of long-chain polyunsaturated fatty acids—both omega-6 and omega-3 types—in *Arabidopsis thaliana*, a type of cress widely used as a model plant in biotechnology research. Dr. Baoxiu Qi and colleagues at the University of Bristol, United Kingdom, transferred to *Arabidopsis thaliana* three genes encoding for different enzymes in the metabolic pathway from linoleic and alphalinolenic acids to arachidonic and eicosapentaenoic acids, respectively. The additional genes were necessary to provide the enzymes to make these long-chain fatty acids. Yields of EPA (13%) and arachidonic acid (29%) in leaves were significantly higher than in conventional cress, which usually does not produce these fatty acids, and accounted for 43% of the total 20-carbon polyunsaturated fatty acids. In addition to the production of EPA and arachidonic acid, the concentration of alpha-linolenic acid was reduced from 48% to 14%. This achievement was also remarkable because it used a pathway seldom found in plants.

This work is important in several regards. One is that it demonstrates the feasibility of developing plants capable of synthesizing long-chain polyunsaturated fatty acids. Another is the relatively high efficiency of conversion of the precursor fatty acids to the long-chain forms. A third advantage is the improved balance of omega-6 and omega-3 fatty acids, with significant reduction in the amounts of the 18-carbon precursors linoleic and alphalinolenic acid compared with conventional plants. Yet another is the demonstration that plants can be engineered not only with respect to the outcome of final products, but also the pathways for achieving the desired ends. A likely next step will be to apply this technology to seed oil crops such as canola and soya bean to see if the long-chain polyunsaturated fatty acids will accumulate in the seed.

Although production of EPA in plants represents an enormous scientific achievement, the question of making Docahexenoic Acid (DHA), a 22-carbon polyunsaturated omega-3 fatty acid important in retina and brain function and other body systems remained unsolved. In mammals, the conversion of EPA to DHA is inefficient and requires several steps. It is possible, in theory, to perform this conversion in a direct manner, but the enzymes to do so are not present in mammals. Several research groups have examined many algae and identified the specific enzymes for this conversion.. Once the genes for these enzymes were identified and cloned they could be incorporated into model organisms to see whether DHA would be produced. In late 2004, Amine Abbadi at the University of Hamburg, Germany, working with colleagues in the U.K. and the U.S., reported the successful transformation of yeast that yielded small amounts of DHA. This accomplishment required four gene transformations. The team then went on to develop transgenic flax, a plant with abundant alpha-linolenic acid for conversion to long-chain fatty acids.

Several steps remain before long-chain polyunsaturated fatty acids will be available in commercial crops. However, the demonstration that plants can be modified to make these important nutrients means that many of the scientific hurdles have been conquered. This work gives a large boost to the potential for plants to be an important dietary source of these fatty acids.

*Gamma-linolenic acid*— This fatty acid is the first step in the conversion of linoleic acid to arachidonic acid in the omega-6 fatty acid pathway. When consumed in evening primrose or borage oils, it is poorly converted to arachidonic acid. For that and other reasons, it may have potential benefit in cardiovascular disease. Gamma-linolenic acid has been associated with improved skin conditions in human subjects, improved liver function in patients with liver cancer, and with anti-cancer effects in cell culture studies. It was also shown to enhance the effectiveness of tamoxifen, an anti-estrogenic medication used to prevent the recurrence of breast cancer. It is believed to suppress the production of estrogen receptors in cells.

Gamma-linolenic acid is naturally present in appreciable amounts in few plants, notably borage (*Borago officinalis*), evening primrose (*Oenothera biennis*), black currant oil (*Ribes nigrum*) and echium (*Echium plantagineum*). The ability to increase the production of gamma-linolenic acid in tobacco plants by transferring the gene for the delta-6 desaturase enzyme from various sources was first shown in 1996 by Reddy and Thomas at Texas A&M University, and by others in 1997. A recent study reported that gamma-linolenic acid content in transgenic canola ranged from 22% to 45%. Evening primrose has also been genetically modified for enhanced gamma-linolenic acid content. Arcadia Biosciences, Davis, CA, has also reported transgenic safflower plants with 65% gamma-linolenic acid in the oil.

Ursin's study of transgenic canola enriched in stearidonic acid discussed above also reported that a cross between the transgenic line of canola producing stearidonic acid and a canola line high in gamma-linolenic acid yielded a canola containing about 11% gammalinolenic acid and about 14% stearidonic acid. This example illustrates the variety of fatty acids that can be developed in seed oils using a combination of genetic engineering and traditional plant breeding techniques.

## Quantity and quality of plant protein

Efforts to improve the protein content and quality of staple foods have been underway for decades. The main focus is crops grown in developing countries, where nutrient shortfalls are widespread and dietary diversity limited. Foods such as potato and cassava, staple foods in several parts of South America and Africa, have less than one percent protein. Efforts to improve protein quality strive to increase the amount of limiting essential amino acids provided by the protein in the food. The amino acids most often present in inadequate amounts are lysine, tryptophan, and methionine. Improvements in protein quality benefit both human and animal nutrition and increase the feed efficiency of crops fed to food animals. For example, corn is widely fed to cattle but it is limiting in lysine and methionine. Corn with higher levels of these amino acids would significantly improve feed efficiency and lower input costs to farmers. Improved corn varieties consumed by humans would also have nutritional benefits.

There are various ways of improving protein quantity and quality. One is to increase the total amount of protein produced by selecting germplasm with an altered balance of seed proteins. This may be done by traditional cross breeding or genetic engineering. Another approach is to introduce genes from other sources for proteins that have a favourable balance of essential amino acids. An example is the introduction into potato of a gene for seed albumin protein from amaranth. A third approach seeks to increase the production of specific amino acids such as lysine. This approach was used in the development of Quality Protein Maize, discussed below. William Folk and his team at the University of Missouri, Columbia, MO, pioneered another approach to improve seed protein quality. Their strategy was to substitute more desirable and scarce amino acids for more abundant ones in certain seed proteins. They applied this concept to rice by increasing the production of lysine, an essential amino acid, at the expense of the non-essential amino acids, glutamine, asparagine and glutamic acid.

*Cassava*: A staple food for some 500 million people in tropical and sub-tropical parts of the world, cassava (*Manihot esculenta* Crantz), also known as yucca or manioc, thrives in marginal lands having little rain and nutrient-poor soils. It is widely consumed in Africa, and parts of Asia and South America. Cassava root has less than 1% protein and poor nutritional value. However, the leaves are also consumed and these are a good source of beta-carotene, the precursor of vitamin A.

In 2003, Zhang and colleagues reported using a synthetic gene to increase the protein content in cassava. The gene is for a storage protein rich in nutritionally essential amino acids. When the gene was expressed in cassava, transformed plants expressed the gene in roots and leaves, both of which are consumed in human diets. The experiment demonstrated the feasibility of increasing the quantity and quality of protein in cassava.

Cassava also contains cyanogenic glucosides that can produce chronic toxicity if not eliminated or reduced by grating, sun-drying, or fermenting. Efforts to develop cassava varieties low in these toxicants is a high research priority. *Corn*: Corn (*Zea mays*) is the predominant staple food in much of Latin America and Africa. Although some varieties may contain appreciable quantities of protein, its quality is poor because of low lysine and tryptophan content. In 1964, it was discovered that corn bearing a gene known as opaque-2 contained increased concentrations of lysine and tryptophan and had significantly improved nutritional quality. However, opaque-2 corn proved to have low yields, increased susceptibility to diseases and pests, and inferior functional characteristics.

At the International Maize and Wheat Improvement Center (CIMMYT) in Mexico, work with the opaque-2 gene continued using both traditional breeding and molecular methods. After at least 12 years' work, CIMMYT researchers succeeded in developing hardy corn varieties that contained twice the lysine and tryptophan content as traditional varieties, but were disease-resistant and high-yielding. Scientists Surinder K. Vasal and Evangelina Villegas of CIMMYT were awarded the World Food Prize in 2000 for their work developing 'Quality Protein Maize'. Quality Protein Maize varieties have been adapted to and released in over 40 countries in Latin America, Africa, and Asia.

Recent researchers at CIMMYT reported the development of transgenic corn with multiple copies of the gene from amaranth (*Amaranthus hypochondriacus*) that encodes for the seed storage protein amarantin. Total protein in the transgenic corn was increased by 32% and some essential amino acids were elevated 8% to 44%. In 2004, a team of researchers at the University of California, Riverside, reported that transgenic corn with increased production of the plant regulating hormone, cytokinin, had nearly twice the content of protein and oil as conventional corn. This development resulted from an unusual change in the way the plant developed. Normally, corn ears develop flowers in pairs, one of which usually dies. Under the influence of the additional cytokinin, both flowers developed but yielded only a single kernel. These kernels contained more protein and oil than conventional corn.

Pursuing a different strategy to improve protein quality, researchers at Monsanto, St. Louis, MO, used genetic engineering techniques to reduce the amount of zein storage proteins. These storage proteins constitute over half the protein in corn and are deficient in lysine and tryptophan. Increased production of other proteins in the corn led to higher levels of lysine, tryptophan, and methionine (Huang et al. 2004). The agronomic and nutritional properties of these lines are currently being evaluated.

Researchers at the Max Planck Institute, Germany, have focused on methionine, another limiting amino acid. They elucidated several key steps in methionine metabolism in plants. This work, currently in the preliminary stage, could pave the way for using genetic engineering techniques to improve the methionine content of plants.

*Potato*— Potato (*Solanum tuberosum*) is a dietary staple throughout parts of Asia, Africa, and South America. Typically, potatoes contain about 2% protein and 0.1% fat. It was reported in 2000 that, as in cassava, transfer of the gene for seed albumin protein from *Amaranthus hypochondriacus* to potato resulted in a 'striking' increase in protein content of the transgenic potatoes. In 2004, researchers at the National Centre for Plant Genome Research, India, reported the development of a nutritionally improved potato line with 25% higher yields of tubers and 35%–45% greater protein content (ISAAA 2004). Dubbed the 'protato,' the protein-rich potato had significant increases in lysine and methionine, which enhance the quality of the additional protein. In February 2004, this potato was reported 'approaching release' to farming communities.

It should be noted that while potatoes are known for their high starch content, it has been possible to genetically engineer potatoes that contain fat (triglycerides). In July 2004, Klaus and colleagues at the Max Planck Institute of Molecular Plant Physiology demonstrated increased fatty acid synthesis in potatoes.

*Rice*—Almost half the world's population eats rice (*Oryza sativa* L.), at least once a day. Rice is the staple food among the world's poor, especially in Asia and parts of Africa and South America. It is the primary source of energy and nutrition for millions. Thus, improving the nutritional quality of rice could potentially improve the nutritional status of nearly half the world's population, particularly its children. Commodity rice contains about 7% protein, but some varieties, notably black rice, contain as much as 8.5% (Food and Agriculture Organization 2004). The most limiting amino acid in rice is lysine. Efforts to increase the nutritional value of rice target protein content and quality along with key nutrients often deficient in rice-eating populations, such as vitamin A and iron.

The International Rice Research Institute (IRRI), Philippines, is a primary center for rice research and development of improved varieties.In 1999, Dr. Momma and colleagues at Kyoto University, Japan, reported a genetically engineered rice having about 20% greater protein content compared with control rice. Transgenic plants containing a soya bean gene for the protein glycinin contained 8.0% protein and an improved essential amino acid profile compared with 6.5% protein in the control rice.

As mentioned briefly above, Dr. William Folk and his team genetically modified rice to increase its content of the amino acid lysine. They did so by modifying the process of protein synthesis, rather than by gene transfer or the expression of new proteins. They achieved an overall 6% increase in lysine content in the grain. Although lysine content remained below optimum levels, the scientists suggested that additional transformations and modifications could further boost lysine levels.

Perhaps the most famous genetic transformations in rice are those in 'Golden Rice' involving the vitamin A precursor, beta-carotene, and iron. The lead scientist in the golden rice project, Dr. Ingo Potrykus, now retired from the Swiss Federal Institute of Technology, was also involved in applying biotechnology for the improvement of rice protein. Although details are sparse, Potrykus described the work of Dr. Jesse Jaynes, who synthesized a synthetic gene coding for an ideal high-quality storage protein with a balanced mixture of amino acids. The gene, named Asp-1, was transferred to rice with the appropriate genetic instructions for its production in the endosperm or starchy part of the rice grain. The transgenic rice plants accumulated the Asp-1 protein in their endosperm in a range of concentrations and provided essential amino acids but data are not yet available on the concentrations achieved or their nutritional relevance. Precedent for the expression of a synthetic gene in rice grown in cell culture suggests that Jaynes' approach is viable.

## Modified Carbohydrate

### Starch

Starch from cereals, grains, and tubers contribute a substantial share of dietary calories and in many poor countries, provide the majority of food energy. Starch is also important for feed and industrial purposes. Its use as paste goes back at least 4000 years BCE to the Egyptians who cemented strips of papyrus stems together with starch paste for writing paper.

Besides providing energy, starch confers functional characteristics to foods: texture, viscosity, solubility, gelatinization, gel stability, clarity, etc. These characteristics depend on the proportion of amylose and amylopectin, the main components of starch. Amylose and amylopectin differ from each other in chain length, branching, and degree of polymerization. Amylose is linear and amylopectin is highly branched. How a particular starch will be used in foods, determines what ratio of amylose to amylopectin is most suitable. High amylose starches include high amylose corn (70%), corn (28%), wheat (26%) and sago (26%). In contrast, waxy rice and waxy sorghum contain no amylose. Members of the potato family—potato, sweet potato, cassava—have 17% to 20% amylose.

Many lines of corn have been developed with different characteristics derived from modified starch ratios and increased amylose content. Transgenic high amylose potatoes developed by inhibiting two branching enzymes were reported to yield more tubers and have lower starch content, smaller granules, and increased reducing sugars. Biotechnology has also been directed to increasing starch content. Potatoes were genetically altered to increase the activity of adenylate kinase, an enzyme involved in the plant's energy metabolism and starch production. The resulting transgenic potatoes had substantially increased adenylates and a 60% increase in starch compared with wild-type plants. Unexpectedly, the concentrations of several amino acids were increased 2- to 4-fold, and tuber yield increased. Considerable publicity was given to potatoes engineered by Monsanto in the early 1990s to have increased starch content. These were touted as more desirable for French fries because they would absorb less fat during frying. They are an example of the type of starch modification that may have secondary health benefits as a consequence of how they are used.

## Fructan

Fructans are polymers (repeating units) of the sugar fructose. They serve in food products as a low-calorie sweetener, source of dietary fiber, and bulking agent. They may also stimulate the growth of desirable colonic bacteria, such as bifida. Fructans have environmentally friendly non-food applications in the manufacture of biodegradable plastics, cosmetics, and detergents. Fructans are naturally occurring in Jerusalem artichokes (sunchokes) and chicory, but agronomic shortcomings in growing these crops have limited their use.

Inulin, a fructan found in Jerusalem artichokes, was successfully synthesized in potatoes following the transfer of two genes from globe artichokes (*Cynara scolymus*). The full spectrum of inulin molecules present in artichokes was expressed in the transgenic potatoes. Inulin comprised 5% of the dry weight of the transgenic tubers and did not influence sucrose concentration. However, starch content was reduced.

In a program called the Agriculture and Fisheries Programme, or FAIR, the European Commission funded multidisciplinary research programs in agriculture and fisheries, including a project on fructans for food and non-food uses. Research to date includes the isolation of several genes for fructosyl transferase enzymes involved in the production of fructans. The feasibility of using these enzymes has been demonstrated in model plants and target crops such as sugar beet. In addition, it was reported in 2004 that genes encoding for fructosyl transferases in onion were isolated and transferred to sugar beet, a plant that does not normally synthesize fructans . Following the transfer of the genes, onion-type fructans were produced from sucrose without loss in storage carbohydrate.

## Increased vitamin content in plants

### A. beta-carotene and other carotenoids

Beta-carotene belongs to the family of carotenoids and is abundant in plants of orange color. It is the precursor of vitamin A and can be converted to the active vitamin during digestion. Other carotenoids do not have potential vitamin A activity. Humans cannot synthesize carotenoids and 'therefore' depend on foods to supply them. Many staple foods, particularly rice, contain no beta-carotene or its precursor carotenoids. Diets lacking other food sources of vitamin A or beta-carotene are associated with vitamin A deficiency which can result in blindness, severe infections, and sometimes death. According to the World Health Organization, vitamin A deficiency is the leading cause of preventable blindness worldwide. The deficiency affects some 134 million people, particularly children, in 118 countries. Overcoming this nutrient deficiency is an urgent global health challenge. The development of 'Golden Rice,' so named because of its yellow colour conferred by the presence of beta-carotene, was a landmark achievement in the application of biotechnology to nutrition and public health. Peter Burkhardt, working with Ingo Potrykus and colleagues in Switzerland, was the first to show that transgenic rice, carrying a gene from daffodil, could express phytoene, a key intermediate in the synthesis of beta-carotene. Subsequently, the Potrykus group reported the application of three transgenes in the development of rice expressing the entire pathway for the production of beta-carotene. Additional work with Golden Rice included the insertion of a gene to increase the iron content. IRRI is currently cross-breeding the nutrient-enhanced transgenic rice with local rice varieties from Asia and Africa, and field testing the new lines for nutritional value and agronomic performance. Varieties of Golden Rice are not expected to be ready for farmers for several more years.

The development of transgenic plants able to produce a variety of carotenoids is an active area of research. It is clear that production of phytoene, the first product in the pathway for carotenoid synthesis, is the rate-limiting step in generating carotenoids. Using gene transfer technology to increase the expression of phytoene synthase, the enzyme that makes phytoene, increases the synthesis of carotenoids substantially. For example, Shewmaker and colleagues (1999) from Monsanto reported an increase up to 50-fold in carotenoids, mainly alpha and beta-carotene, in canola. However, vitamin E levels decreased significantly, oleic acid content increased, and linoleic and alpha-linolenic acids were reduced compared with non-transgenic seeds. These other changes would have to be modified or evaluated to determine whether they might have meaningful nutrition implications.

In a separate study on canola, the Monsanto group transferred to canola three genes from bacteria that affect the phytoene synthesis pathway. When they included a triple construct— genes for three different enzymes, phytoene synthase, phytoene desaturase and lycopene cyclase—the resulting transgenic canola seeds maintained the same amount of total carotenoids, but increased the ratio of beta to alpha-carotene from 2:1 to 3:1.

Stalberg and colleagues in Sweden also studied the effects of phytoene synthase on carotenoid synthesis in transgenic *Arabidopsis thaliana*. They examined three ketocarotenoids; transformed seeds had a 4.6-fold increase in total pigment and a 13-fold increase in these three carotenoids. They also reported a 43-fold average increase in beta-carotene. Lutein, another nutritionally important carotenoid, was significantly increased, but zeaxanthin was only increased by a factor of 1.1. They also observed substantial levels of lycopene and alpha-carotene in the seeds, whereas only trace amounts were found in the control plants. However, germination was delayed in proportion to the increased levels of carotenoids.

Others have examined the effect of transgenes affecting phytoene metabolism on carotenoid synthesis. Dr. Peter Bramley's group at the University of London, United Kingdom, transformed tomatoes using a bacterial gene encoding for an enzyme that converts phytoene to lycopene, the precursor of beta-carotene. Tomatoes carrying the bacterial gene had about a 3-fold increase in beta-carotene content, but total carotenoids were not increased. The altered carotene content did not affect plant growth and development.

Lutein and zeaxanthin are nutritionally important carotenoids for protection of the retina and reduced risk of age-related macular degeneration. Lutein is found in dark green leafy vegetables such as spinach and collards, and zeaxanthin occurs in yellow foods such as mangoes, corn, and peaches. The latter is not particularly abundant in Western diets. Romer and colleagues at Universitat Konstanz, Germany, were able to use biotechnology to block the conversion of zeaxanthin to another carotenoid and thereby increase its content in potatoes. With this approach they obtained increased levels of zeaxanthin in potatoes ranging from 4- to 130-fold. Total carotenoids were increased by 5.7-fold, but in some, lutein content was decreased. Alphatocopherol (vitamin E) was increased 2- to 3-fold in the transgenic potatoes. Fine-tuning these alterations has the potential to significantly enhance the nutritional value of potatoes.

In another study, Bramley's group transferred the gene that increases carotenoid synthesis from a bacterium to tomatoes and measured total and specific carotenoids in the transgenic fruits. Total carotenoids were 2- to 4-fold higher in transgenic fruits than in nontransformed plants, with increases in phytoene, lycopene, beta-carotene, and lutein of ranging from 1.8- to 2.4-fold. Tomatoes with delayed ripening were produced as a result of inserting a gene encoding for S-adenosylmethionine decarboxylase, an enzyme involved in the ripening process. An additional consequence of this transgenic modification was a severalfold increase in lycopene content. Lycopene is normally converted to beta-carotene, but tomatoes with increased lycopene content may have enhanced nutritional value. Lycopene consumption has been linked to reduced risk and spread of prostate cancer, though definitive data are lacking.

The carotenoid astaxanthin is synthesized by algae and plants and is responsible for the pink colour in shrimp and salmon. Humans absorb astaxanthin poorly, but absorption is increased in the presence of fat. Astaxanthin is of interest because of its strong antioxidant properties *in vitro*. It is less certain whether it is an antioxidant in human health. Astaxanthin is used commercially in feed for cultured salmon and trout.

Production of astaxanthin in flowers and fruits has also been accomplished with the techniques of biotechnology. Using a gene from the alga *Haematococcus pluvialis*, researchers at The Hebrew University of Jerusalem, Israel, transferred the gene into tobacco (*Nicotiana tabacum*). Transgenic tobacco plants produced astaxanthin and changed colour. This ability to manipulate pigmentation in fruits and flowers may have commercial potential and possible implications for increasing the availability of carotenoids for human health.

## B. Vitamin E

There is strong interest in vitamin E because of studies linking it to decreased occurrence of several degenerative diseases and cancers, although efficacy remains unproven and data are inconsistent. Some recent trials with vitamin E supplementation reported no protection against cardiovascular disease or cancer and some chance of increased risk of heart failure. Also, because vitamin E is an anti-oxidant, it is useful in foods and oils to provide oxidative stability. Vitamin E is found mainly in vegetable oils, wheat germ, and a few other foods not widely consumed. The most potent form of the vitamin is alpha-tocopherol, but the less potent gamma, beta, and delta forms are more widespread in plants. Efforts to increase the content of vitamin E in food plants, particularly cereals and grains, which may have low amounts, have sought to increase the amount of precursor substances by overexpressing the genes for various enzymes involved in the biosynthetic pathway. Vitamin E biosynthesis involves complicated pathways so that multiple genetic manipulations are required.

In September 2003, Dr. Edgar Cahoon of the U.S. Department of Agriculture (USDA) and co-researchers at the Donald Danforth Plant Science Center, St. Louis, MO, announced the development of transgenic corn with increased levels of vitamin E. Insertion of a gene from barley into corn increased the conversion of vitamin E precursors to vitamin E itself. The content of vitamin E and tocotrienol, a closely related substance, was increased up to 6-fold. However, much of the antioxidant produced was tocotrienols rather than vitamin E. Tocotrienols, although potent antioxidants *in vitro*, are poorly absorbed in humans; however, they may have cholesterol lowering properties. Besides enhancing the potential therapeutic and nutritional value of corn, this alteration may increase its oxidative stability after harvest. In soya beans, Van Eenennaam and colleagues developed transgenic plants that were able to increase the conversion of the weaker forms of tocopherol typically present in soya beans to the more potent alpha-tocopherol (vitamin E) form. The result was a 5-fold increase in vitamin E activity. This work paves the way for the development of vitamin E-rich oils and plants with potential health benefits.

### C. Vitamin C

Vitamin C, or ascorbic acid, is abundant in citrus and other fruits such as strawberry and kiwi, but is very low in cereals and grains. Moreover, it cannot be synthesized by humans nor stored to any appreciable extent. Thus, we depend on regular dietary consumption to meet our vitamin C needs. In areas of the world where foods containing vitamin C are not widely consumed, strategies to increase the vitamin C in cereals and grains hold considerable potential to improve health.

In 2003, Gallie and colleagues at the University of California, Riverside, announced the successful transformation of corn and tobacco that resulted in a 20% increase in vitamin C. They accomplished this increase by transferring a gene from wheat for an enzyme that recycles vitamin C and prevents its breakdown. Increased expression of this enzyme in transgenic corn and transgenic tobacco resulted in a 2- to 4-fold increase in vitamin C content in the kernel. Application of this approach to other food plants remains to be developed and evaluated.

The ability to increase the vitamin C content of strawberries, a good source of the vitamin, was reported in 2003 by a research team at the University of Malaga, Spain. Using a gene for the enzyme D-galacturonate reductase transferred to strawberry (*Fragaria spp*) this group showed that vitamin C content in the modified strawberry fruit increased with the expression of the transgene. This study demonstrated the feasibility of using this enzyme to raise the level of vitamin C in plants.

### D. Folic Acid

Inadequate intake of the vitamin folic acid, one of a family of folates, is associated with megaloblastic anemia, birth defects, impaired cognitive development, and some chronic diseases. Folates are available in small amounts in a variety of fruits and vegetables, but intakes tend to be low. For this reason, it would be desirable to increase the concentration of folates in dietary staples and foods widely consumed. In the U. S. several foods are fortified with folic acid. In 2004, Hossain and colleagues at Tufts University, Medford, MA, and the Donald Danforth Plant Science Center reported a 2- to 4-fold increase in folates and pterins, the precursors of folates, in transgenic *Arabidopsis thaliana* modified by the incorporation of the transgene for the first step in the synthesis of folic acid. Other investigators have developed transgenic tomatoes, also enriched in the same gene, that had twice the amount of folate as control fruit. This group was able to boost folate content 10-fold by including a second gene transformation to increase the content of another substance, para-aminobenzoate, needed for folate synthesis. These studies provide good evidence of the potential to increase the availability of this vitamin in widely consumed foods.

## E. Antioxidants

Vitamins C and E function as antioxidants in the body. However, many other substances widely distributed in foods in small amounts also provide protection against potentially damaging breakdown products arising from oxidation during normal metabolism. Oxidation breakdown products, such as reactive oxygen species, oxidized lipids, and free radicals have been associated with chronic diseases, so there has been considerable interest in the availability of antioxidants. Caution should be sounded, however, because in small amounts many of these substances are protective; in high doses, they can act as prooxidants and may be harmful. Several examples of the applications of biotechnology for enhanced antioxidant capacity in foods are described below.

Phenolic compounds are the most widespread antioxidants in foods. They include such substances as flavanols, tocopherols, quercetin, resveratrol, and many others. They have become familiar to consumers because they are touted in foods as diverse as berries, wine, tea, olive oil, and many others. Potatoes are a source of antioxidant flavanoids and vitamin C. To enhance the antioxidant content of potatoes, Lukaszewicz and colleagues conducted a series of transformations using one or multiple genes encoding enzymes in the bioflavanoid synthesis pathway. Transgenic plants exhibited significantly increased levels of phenolic acids and anthocyanins, plus improved antioxidant capacity. However, starch and glucose levels were decreased. These findings point to complex relationships between antioxidant content and other compounds, but indicate that antioxidant levels in potatoes can be altered using biotechnology.

Another phenolic antioxidant, chlorogenic acid, accumulates in some crops and is found in apples, green coffee beans, tomatoes, and tea. It is synthesized by the enzyme hydroxycinnamoyl transferase in solanaceous plants (e.g., potato, tomato, eggplant). In 2004, it was reported that transgenic tomatoes carrying the gene for this enzyme accumulated higher levels of chlorogenic acid with no side effects on levels of other phenolics. The transgenic tomatoes also showed improved antioxidant capacity, suggesting that such enhanced tomatoes might provide additional antioxidants.

Yet another transformation in tomatoes was recently reported to result in the synthesis of resveratrol, an antioxidant not normally found in tomatoes. Resveratrol is usually associated with grapes and wines where it is abundant. In this study, tomatoes incorporating the gene for stilbene synthase, an enzyme in the pathway for resveratrol synthesis, had a resveratrol content of 53 mg/g fresh tomato upon ripening. The contents of two other antioxidants, vitamin C and glutathione, were also increased.

## VI. Trace mineral content and bioavailability

Improving human nutrition by increasing the availability of trace minerals in crops is potentially highly efficient and effective. This strategy may reach more people in developing countries than fortification of foods, because many subsistence farmers grow their own food and are outside the market system. If they have access to and consume improved crop varieties, they will not only improve their nutrient intake, they may improve their crop yields and consequently their economic wellbeing. This is because trace minerals are essential to the plant's ability to resist disease and other environmental stresses. Further, plants with improved ability to take up minerals from the soil will not deplete nutrient-poor soils. Such plants are able to unbind minerals in the soil and make them available to the plant, thus making use of an abundant resource in the soil that is otherwise unavailable. Mineral-efficient plants are also more drought resistant and require fewer chemical inputs.

### A. Iron

Iron deficiency anemia is one of the most widespread nutritional deficiencies in the world. The United Nations estimates that over three billion people in developing countries are iron-deficient. The problem for women and children is more severe because of their greater need for iron. For this reason, the enrichment of staple foods, especially those consumed in poor countries, is one of the top priorities in international agricultural and nutrition research. In rice-eating populations, iron deficiency anemia is caused by insufficient dietary iron, absorption inhibitors such as phytate, and lack of enhancing factors for iron absorption such as ascorbic acid.

Although much is known about the uptake of iron and zinc in roots and transport of minerals to and from vegetative parts of the plant, some plants accumulate very little trace minerals in the grain. For example, in wheat only 20%, and in rice just 5% of the iron in leaves is transported to the grain. In cereals, much is stored in the husk and subsequently lost during milling and polishing. Thus, strategies to increase the iron content of cereals and grains face the challenge of targeting iron storage in a form and location in the plant where it will be bioavailable when consumed. A significant breakthrough in improving the iron content of cereals was achieved by Ingo Potrykus and colleagues. One of the genes transferred to Golden Rice came from the common bean, *Phaseolus vulgaris*. This gene encoded for the iron storage protein, ferritin, and when expressed in the transgenic rice increased the iron content twofold. The bioavailability of iron in transgenic rice varieties containing ferritin was shown to be as good as ferrous sulphate, commonly used in iron supplements, as reflected in biochemical indices of iron status in iron-deficient laboratory rats.

A different source of ferritin genes, soya bean, was used in the transformation of rice to increase iron content. Researchers at the Central Research Institute Electric Power Industry, Japan, transferred the gene for ferritin from soya bean into rice and confirmed the stable incorporation of the ferritin subunit in the rice seed. Iron content in the transgenic rice seeds was up to threefold greater than in non-transgenic control plants. Others have used recombinant soya bean ferritin under a different promotor to increase the iron content in wheat and rice. In this case, iron was significantly increased in vegetative tissues but not in seeds. Thus, the experimental conditions, type of promotor used, mineral transport and storage in the plant, and other conditions have substantial effects on the outcome of genetic engineering experiments to increase mineral content.

Iron transport and uptake in plants is carefully regulated. This is because iron has low solubility and is toxic in excess. Recent studies have examined the function of iron transporter proteins in transgenic plants. These proteins have been shown to increase iron uptake into roots when iron is deficient. The iron transporter protein, IRT1, first isolated from *Arabidopsis thaliana*, also transports other metals such as zinc, manganese, lead, and cadmium; the latter two can be toxic. Researchers in the laboratory of Dr. Mary Lou Guerinot at Dartmouth College, Hanover, NH, have shown that slight changes in the amino acid composition of the transporter protein affects the selectivity of metals transported into the plant. This finding introduces the possibility of engineering plants with the ability to take up desirable minerals while excluding toxic and undesirable ones.

A second iron transporter protein, IRT2, has also been identified in the roots of *Arabidopsis*. When the gene for IRT2 was incorporated into iron- and zinc-deficient plants, iron uptake was increased. Unlike the IRT1 transporter, however, IRT2 did not transport manganese and cadmium when it was expressed in yeast. This observation suggests ways in which selective genetic transformations might be used to enhance the uptake of some minerals while excluding others.

It should be noted that iron and zinc levels tend to be present together in many plants, although the average content of each differs. For example, in screening over 1,000 varieties of common beans, a nutritional staple in many countries, scientists at the International Institute of Tropical Agriculture, Nigeria, found that iron content averaged about 55 mg/ g iron, but some varieties from Peru averaged more than 100 mg/g iron.

Zinc content, averaged 35 mg/g. When varieties were selected for their iron content, higher zinc levels were obtained as well. These observations suggest that genetic modification of selected varieties to further increase iron content might boost zinc levels too. As in beans, iron and zinc concentrations differ across varieties of rice. Aromatic rice tends to have the highest iron levels and several varieties have been successfully crossed with elite rice lines having excellent agronomic characteristics and grain qualities. These micronutrient traits were shown to be stable across different growing environments and could be crossed with high yielding varieties to improve the nutrient density. High iron rice developed from traditional breeding is currently being tested for iron bioavailability and effects on iron nutrition status in young women in the Philippines. Results are not yet available. Another approach to improving the availability of iron for infants was reported by Suzuki and colleagues at the University of California, Davis. These investigators developed transgenic rice in cell culture that expressed the gene for human lactoferrin, a milk protein that binds iron. When they compared the recombinant lactoferrin with native human lactoferrin both proteins retained functional activity after mild heat treatment, high acidity, and *in vitro* digestion. Their findings suggest that recombinant lactoferrin grown in plant culture may be a functional alternative to human lactoferrin in infant formula and provide another way to improve iron availability during infancy.

## B. Zinc

Zinc is one of several trace minerals that can be deficient in human diets, especially where meat is not consumed. Zinc deficiency is associated with impaired growth and reproduction, anorexia, immune disorders, and a variety of other symptoms. Zinc is also an important constituent of more than 100 enzymes. Absorption of zinc from cereals and grains can be impaired or blocked by the presence of some substances such as phytate. Increasing the zinc content of cereals and grains, especially where soils are low in zinc, may be an effective way to improve human nutrition and at the same time increase plant yields. Ramesh and colleagues in Australia, studied the effect on seed zinc content in barley (*Hordeum vulgare* cv Golden Promise) in plants transformed to increase the expression of zinc transporter enzymes. Multiple transgenic lines exhibited higher zinc and iron contents in their seeds compared with control plants. When grown under zinc deficient conditions, zinc uptake in the transgenic lines was higher in the short term compared with control plants. When zinc content was restored, uptake of zinc decreased in both transgenic and control plants, suggesting that the transporter proteins may be degraded when zinc is adequate. This study suggests that increasing the production zinc transporter proteins may be one approach to increasing the zinc content of cereals.

### C. Selenium

Selenium is an essential trace mineral incorporated into plants from soil. Consumption of selenium has been linked to reduced risk of all cancers, but particularly those of the lung, colo-rectum and prostate. Selenium is also important for specific enzymes and proteins in the brain and is necessary for proper immune function. However, selenium is toxic at levels only a little greater than those required in a healthy diet, so caution is warranted with supplementation and increased intakes. Areas where soils are deficient in selenium are well known and low to deficient intakes have been observed among human and animal populations in these regions. Genetic engineering technology offers considerable potential for increasing the uptake of selenium from soils and incorporating the mineral into non-toxic compounds in the edible parts of plants.

CHAPTER-8

# The Looming Trade War Over Plant Biotechnology

The battle lines are being drawn. On one side stands the United States, the world's leading developer and exporter of genetically modified crops. On the other is the European Union, whose consumers, spooked by anti-biotechnology activists, are demanding that all biotech crops be labeled if not banned altogether. Caught in the middle are American farmers, who plant more than two-thirds of all the world's acreage devoted to genetically enhanced crops. The U.S. Department of Agriculture estimates that genetically modified crops will represent nearly one-third of the 2002 corn harvest and nearly three-quarters of the 2002 U.S. cotton and soya bean harvests. The EU is one of the most important potential markets for those crops. American farmers already export about 30 percent of their soya bean harvest and 20 percent of their corn harvest to the EU. American farmers exported $6.3 billion in agricultural goods to the EU in 2000. Twenty-four percent of those exports were oilseed products, chiefly soya beans and soya products, and 16 percent were grains and feeds. Sixty-three percent of U.S. corn byproduct exports went to the EU.1 U.S. corn growers alone have lost about $200 million per year since 1998 because of the EU ban on importing genetically enhanced crops.2 Impending

EU regulations on biotech crops would seriously disrupt the flow of those exports to European markets.

The Office of the U.S. Trade Representative has already threatened to bring the issue to the World Trade Organization in Geneva for adjudication. This transatlantic food fight has broader implications as well. If the U.S. position prevails, the poor of the world will have access to a safe technology that could dramatically reduce hunger and malnutrition. If the EU position prevails, research will slow, putting the world's poor at greater risk of starvation and setting a terrible precedent for the future of free trade.

This analysis will answer four questions: (1) What is plant biotechnology? (2) Who opposes it and why? (3) Where does the trade battle stand now? (4) What should U.S. policy be?

## What is plant biotechnology?

In the last decade, biologists and crop breeders have made enormous strides in their ability to select specific useful genes from various species and splice them into unrelated species. Previously, plant breeders were limited to introducing new genes through the time consuming and inexact art of crossbreeding species that were fairly close relatives, for example, rye and wheat, plums and apricots. For each cross, thousands of unwanted genes would necessarily be introduced into a crop variety. Years of 'backcrossing'—breeding each new generation of hybrids with the original commercial variety over several generations— were needed to eliminate the unwanted genes so chiefly useful genes and characteristics remained. The new biotech methods are far more precise and efficient. The plants they produce are variously described as 'transgenic,' 'genetically modified,' 'genetically engineered,' or 'genetically enhanced.' Plant breeders using biotechnology have accomplished a great deal in only a few years. For example, they have created a class of highly successful insect-resistant crops by incorporating toxin genes from the soil bacterium *Bacillus thuringiensis*. Farmers have sprayed *B. thuringiensis* spores on crops as an effective insecticide for decades. Now, thanks to some clever biotechnology, breeders have produced varieties of corn, cotton, and potatoes that make their own insecticide. *B. thuringiensis* is toxic largely to destructive caterpillars such as the European corn borer and the cotton bollworm; it is not harmful to birds, fish, mammals, or people. Another popular class of biotech crops incorporates an herbicide-resistance gene that has been especially useful in soya beans. Farmers can spray herbicide on their fields to kill weeds without harming the crop plants. The most widely used herbicide is Monsanto's Roundup (glyphosate), which toxicologists regard as an environmentally benign chemical that degrades rapidly, only days after being applied. Farmers who use 'Roundup Ready' crops don't have to plow for weed control, which means there is far less soil erosion.

Biotech is the most rapidly adopted new farming technology in history. The International Institute for the Acquisition of Agri-Biotech Applications estimates that the global area planted in biotech crops in 2001 was 130 million acres (52.6 million hectares), up 19 percent from 2000. The area planted in biotech crop varieties is up 30-fold since 1996. The first generation of biotech crops was approved by the Environmental Protection Agency, the Food and Drug Administration, and the U.S. Department of Agriculture in 1995. The USDA estimates that in 2002 transgenic varieties will account for 32 percent of corn acreage, 74 percent of soya bean acreage, and 71 percent of cotton acreage in the United States. With biotech soya beans, U.S. farmers save an estimated $216 million annually in weed control costs and make 19 million fewer herbicide applications per year. In addition, using no-till farming made possible by herbicide- resistant biotech soya beans, farmers prevent 247 million tons of topsoil from eroding away. It is estimated that herbicide-resistant biotech soya beans, canola, cotton, and corn varieties and insect-resistant biotech cotton reduced global pesticide use by 22.3 million kilograms of formulated product in 2000. U.S. cotton farmers avoided spraying 2.7 million pounds of insecticides and made 15 million fewer pesticide applications per year by switching to biotech varieties. Their net revenues increased by $99 million.10 Researchers estimate that *B. thuringiensis* corn, by preventing insect damage, increased yields by 66 million bushels in 1999.

### Documented safety

One scientific panel after another has concluded that biotech foods are safe to eat, and so has the FDA. Since 1995, tens of millions of Americans have been eating biotech crops. Today, it is estimated that 60 percent of the foods on American grocery shelves are produced using ingredients from transgenic crops. In April 2000 a National Research Council panel issued a report that emphasized that the panel could not find 'any evidence suggesting that foods on the market today are unsafe to eat as a result of genetic modification.'

*Transgenic Plants and World Agriculture*, a 2000 report prepared under the auspices of seven scientific academies in the United States and other countries, strongly endorsed crop biotechnology, especially for poor farmers in the developing world. 'To date,' the report concluded, "Over 30 million hectares of transgenic crops have been grown and no human health problems associated specifically with the ingestion of transgenic crops or their products have been identified."

Both reports concurred that genetic engineering poses no more risks to human health or to the natural environment than does conventi onal plant breeding. As biologist Martina McGloughlin of the University of California at Davis remarked at a Congressional Hunger Center seminar in June 2000, "The biotech foods on our plates have been put through more thorough testing than conventional food ever has been subjected to."

According to a report issued in April 2000 by the House Subcommittee on Basic Research: "No product of conventional plant breeding . . . could meet the data requirements imposed on biotechnology products by U.S. regulatory agencies. . . . Yet, these foods are widely and properly regarded as safe and beneficial by plant developers, regulators, and consumers." The report concluded that biotech crops are at least as safe as and probably safer than conventionally bred crops. Even a 2001 review of 81 separate European scientific studies of genetically modified organisms funded by the European Union found no evidence that genetically modified foods posed any new risks to human health or the environment.

## Feeding the hungry of the world

Today, pest resistance and herbicide resistance, along with some disease resistance traits, are the chief improvements incorporated into biotech crops. And most of those enhancements have been made in leading commercial crops, such as corn, soya beans, and cotton, grown in developed countries. The next frontier will be applying genetic enhancements to crops that will feed the hungry in developing countries. However, progress could be halted if a full-fledged trade war breaks out between the United States and the EU, increasing the risk of starvation for millions. The International Food Policy Research Institute estimates that global food production must increase by

40 percent in the next 20 years to meet the goal of a better and more varied diet for a world population of some 8 billion people. As biologist Richard Flavell concluded in a 1999 report to the IFPRI, "It would be unethical to condemn future generations to hunger by refusing to develop and apply a technology that can build on what our forefathers provided and can help produce adequate food for a world with almost 2 billion more people by 2020."

The good news is that researchers are already at work on improving crops that will help the poor in developing countries. For example, researchers have developed 'golden rice,' a crop that could prevent blindness in from .5 million to 3 million poor children a year and alleviate vitamin A deficiency in some 250 million people in the developing world. By inserting three genes, two from daffodils and one from a bacterium, scientists at the Swiss Federal Institute of Technology created a variety of rice that produces the nutrient beta carotene, the precursor to vitamin A. Agronomists at the International Rice Research Institute in the Philippines plan to crossbreed the variety, called 'golden rice' because of the color produced by the beta carotene, with well-adapted local varieties and distribute the resulting plants to farmers all over the developing world.

Technologies already well understood in the developed world are also valuable for farmers in the developing world. Thousands of poor Indian farmers nearly rioted in early 2002 when the Indian government, spurred by antibiotech activists, seemed poised to destroy the biotech pest-resistant cotton the farmers had planted. Faced with a possible farmer revolt, the Indian government backed down and approved the biotech cotton for planting. Another way biotech crops can help poor farmers grow more food is by controlling parasitic weeds, an enormous problem in tropical countries. Cultivation cannot get rid of them, and farmers must abandon fields infested with them after a few growing seasons. Herbicide resistant crops, which would make it possible to kill the weeds without damaging the cultivated plants, would be a great boon to such farmers.

Kenyan biologist Florence Wambugu argues that crop biotechnology has great potential to increase agricultural productivity in Africa without demanding big changes in local practices. A drought-tolerant seed will benefit farmers whether they live in Kansas or Kenya.

## Fighting drought and plant diseases

By incorporating genes for proteins from viruses and bacteria, crops can be immunized against infectious diseases. The papaya mosaic virus had wiped out papaya farmers in Hawaii, but a new biotech variety of papaya incorporating a protein from the virus is immune to the disease. As a result, Hawaiian papaya orchards are producing again, and the virus-resistant variety is being made available to developing countries. Similarly, scientists at the Donald Danforth Plant Science Center in St. Louis are at work on a cassava variety that is immune to cassava mosaic virus, which killed half of Africa's cassava crop two years ago. Biotech companies are granting to international and academic research institutes broad licenses to use their patents. That will enable the development of genetically enhanced crops, such as cassava and rice, that are especially important to poor farmers in the developing world.

Another recent advance with enormous potential is the development of biotech crops that can thrive in acidic soils, a large proportion of which are located in the tropics. Aluminum toxicity in acidic soils reduces crop productivity by as much as 80 percent. Progress is even being made toward the Holy Grail of plant breeding, transferring the ability to fix nitrogen from legumes to grains. (Legumes such as soya beans and alfalfa house microorganisms in their roots that allow them to absorb nitrogen from the atmosphere and transform it into bio-logically useful forms—that is, literally make nitrogen fertilizer, which all plants need, using their roots.) That achievement would greatly reduce the need for fertilizer. Biotech crops with genes for drought and salinity tolerance are also being developed. Researchers at the University of California at San Diego have already identified techniques that could make plants more drought resistant. McGloughlin predicts, Further down the road, we will be able to use biotechnology to enhance nutritional content of crops such as protein, vitamins, minerals, and antioxidants, remove anti-nutrients, remove allergens, and remove toxins. We will also be able to enhance other characteristics such as growing seasons, stress tolerance, yields, geographic distribution, disease resistance, shelf life and other properties of production of crops. The ability to manipulate plant nutritional content heralds an exciting new area and has the potential to directly benefit developing countries."

Biotech crops can provide medicine as well as. food. Biologists at the Boyce Thompson Institute for Plant Research at Cornell University recently reported success in preliminary tests with biotech potatoes that would immunize people against diseases. One modification protects against Norwalk virus, which causes diarrhea, and another might protect against the hepatitis B virus, which afflicts 2 billion people. Plant-based vaccines would be especially useful for poor countries, which could manufacture and distribute medicines grown by local farmers.

Plant biotechnology has already dramatically boosted American farmers' productivity and lowered their costs and, at the same time, helped them to protect the natural environment by reducing their use of agricultural chemicals and preventing soil erosion. Consumers also benefit from lower prices and a healthier environment. In the future consumers will benefit even more as biotechnologists develop fresher and more nutritious foods along with crop- and plant-derived medicines and vaccines. A robust plant biotechnology industry coupled with American farming prowess will also ensure that our country remains the granary to the world. In developing countries, the deployment of plant biotechnology can spell the difference between life and death and between health and disease for hundreds of millions of the world's poorest people.

## Who opposes plant biotechnology and why?

There is a growing global war against crop biotechnology. Gangs of anti-biotech vandals with cute monikers such as Cropatistas and Seeds of Resistance have ripped up scores of research plots in Europe and the United States. The socalled Earth Liberation Front burned down a crop biotech lab at Michigan State University on New Year's Eve in 1999, destroying years of work and causing $400,000 in property damage.

Overall, the Federal Bureau of Investigation estimates that ELF has perpetrated more than 600 attacks and caused $43 million in damage since 1996. Anti-biotech lobbying groups have proliferated and now include Greenpeace, the Union of Concerned Scientists, the Institute for Agriculture and Trade Policy, the Institute of Science in Society, the ETC (Action Group on Erosion, Technology and Concentration) Group, the Ralph Nader–founded Public Citizen, the Council for Responsible Genetics, the Institute for Food and Development Policy, and that venerable opponent of technological change, Jeremy Rifkin's

Foundation on Economic Trends.

## False alarms

Despite the wide agreement among scientific and medical organizations on the safety of biotech crops, activists still insist that those crops are not safe. For example, they point to a study by Arpad Pusztai, a researcher at Scotland's Rowett Research Institute, that was published in the British medical journal the *Lancet* in October 1999. Pusztai found that rats fed one type of genetically modified potatoes (not a variety created for commercial use) developed immune system disorders and organ damage.

The *Lancet*'s editors, who published the study even though two of six reviewers rejected it, apparently were anxious to avoid the charge that they were muzzling a prominent biotech critic. But the *Lancet* also published a thorough critique, which concluded that Pusztai's experiments were incomplete, included too few animals per diet group, and lacked controls such as a standard rodent diet. . . . Therefore the results are difficult to interpret and do not allow the conclusion that the genetic modification of potatoes accounts for adverse effects in animals.

The Rowett Institute, which does mainly nutritional research, fired Pusztai on the grounds that he had publicized his results before they had been peer reviewed. Activists are also fond of noting that the seed company Pioneer Hi-Bred produced a soya bean variety that incorporated a gene—from a protein in Brazil nuts—that causes reactions in people who are allergic to nuts. The activists fail to mention that the soya bean never got close to commercial release because Pioneer Hi-Bred checked it for allergenicity as part of its regular safety testing and immediately dropped the variety. The other side of the allergy coin is that biotech can remove allergens that naturally occur in foods such as nuts, potatoes, and tomatoes, making those foods safer. In October 2000 activists seized on the news that a genetically modified corn variety called StarLink that was approved only for animal feed in the United States had been inadvertently used in two brands of taco shells, prompting recalls and front-page headlines. Ultimately, compensating food companies and growers for the recall cost Aventis, the creator of StarLink, $1 billion.

Lost in the furor was the fact that there was little reason to believe the corn was unsafe for human consumption—only an implausible, unsubstantiated fear that it might cause allergic reactions. The U.S. Centers for Disease Control found that there was no evidence that anyone had suffered any adverse reaction to eating foods containing StarLink corn.

## Butterfly friendly

Activists also cite environmental concerns as a reason to oppose plant biotechnology. Most notoriously, activists worry about how biotech corn pollen affects the monarch butterfly. The global campaign against green biotech received a public relations windfall on May 20, 1999, when *Nature* published a study by Cornell University researcher John Losey that found that monarch butterfly caterpillars died when force-fed milkweed dusted with pollen from *B. thuringiensis* corn. Since then, at every anti-biotech demonstration, the public has been treated to flocks of activist women dressed as monarch butterflies. But when more-realistic field studies were conducted, researchers found that 'there is no significant risk to monarch butterflies from environmental exposure to Bt corn.' Corn pollen is heavy and doesn't spread very far, and milkweed grows in many places in addition to the margins of cornfields. In the wild, monarch aterpillars apparently know better than to eat corn pollen on milkweed leaves.

Furthermore, *B. thuringiensis* crops mean that farmers don't have to indiscriminately spray their fields with insecticides, which kill beneficial as well as harmful insects. In fact, studies show that *B. thuringiensis* cornfields harbour higher numbers of beneficial insects such as lacewings and ladybugs than do conventional cornfields. James Cook, a biologist at Washington State University, points out that the population of monarch butterflies has been increasing in recent years, precisely the time period in which *B. thuringiensis* corn has been widely planted. The fact is that pest-resistant crops are harmful mainly to target species—that is, exactly those insects that insist on eating them. Never mind; we will see monarchs on parade for a long time to come. Meanwhile, a spooked Environmental Protection Agency has changed its rules governing the planting of *B. thuringiensis* corn, requiring farmers to plant non–*B. thuringiensis* corn near the borders of their fields so that *B. thuringiensis* pollen doesn't fall on any milkweed growing there. But even the EPA firmly rejects activist claims about the alleged harms caused by *B. thuringiensis* crops.

Prior to registration of the first *B.t.* plant pesticides in 1995, it said in response to a Greenpeace lawsuit, "EPA evaluated studies of potential effects on a wide variety of non-target organisms that might be exposed to the *B.t.* toxin, e.g., birds, fish, honeybees, ladybugs, lacewings, and earthworms." The EPA concluded, "These risk assessments demonstrated that Bt endotoxins expressed in transgenic plants do not exhibit detrimental effects to nontarget organisms in populations exposed to the levels of endotoxin found in plant tissue."

In other words, those species were not harmed by transgenic plants. In a review article in *Nature Biotechnology*, researchers strongly concurred: "In most cases, no adverse effects were observed even when test populations were exposed to [bt] toxin concentrations over 500-1,000-fold greater than those they would be expected to encounter under field conditions."

## Runaway crossbreeding and superpests?

Another danger highlighted by anti-biotech activists is the possibility that transgenic crops will crossbreed with other plants. At the Congressional Hunger Center seminar, British activist Mae-Wan Ho claimed that genetically modified constructs 'are designed to invade genomes and to overcome natural species barriers.' And that's not all. "Because of their highly mixed origins," she added, "GM constructs tend to be unstable as well as invasive, and may be more likely to spread by horizontal gene transfer." In other words, genetically modified organisms (GMOs) could supposedly spawn new and harmful breeds unintended by their creators.

"Nonsense," says Tuskegee University biologist C. S. Prakash. "there is no scientific evidence at all for Ho's claims." Prakash points out that plant breeders specifically choose transgenic varieties that are highly stable since they want the genes that they've gone to the trouble and expense of introducing into a crop to stay there and do their work. Ho also charges that GM genetic material, when eaten, is far more likely to be taken up by human cells and bacteria than is natural genetic material. Again, there is no scientific evidence for this claim. All genes from whatever food sources are made up of the same four DNA bases, and all undergo digestive degradation when eaten. Britain's chief scientific organization, the Royal Society, issued a report in February 2002 that pointed out this elementary fact of biology when it concluded, "Given the very long history of DNA consumption from a wide variety of sources, it is likely that such consumption poses no significant risk to human health, and that additional ingestion of GM DNA has no effect."40 Opponents of biotech also sketch scenarios in which transgenic crops foster 'superpests': weeds bolstered by transgenes for herbicide resistance or pesticide-proof bugs that proliferate in response to crops with enhanced chemical defenses. As McGloughlin notes: "The risk of gene flow is not specific to biotechnology. It applies equally well to herbicide-resistant plants that have been developed through traditional breeding techniques." Even if a herbicide resistance gene did get into a weed species, most researchers agree that it would be unlikely to persist unless the weed were subjected to significant and continuing selection pressure—that is, was sprayed regularly with a specific herbicide. And if a weed becomes resistant to one herbicide, it can be killed by another.

Conventional spray pesticides encourage the evolution of pesticide-resistant insects, so there is no scientific reason for singling out biotech plants. Cook points out that crop scientists could handle growing pesticide resistance the same way they deal with resistance to infectious rusts in grains: using conventional breeding techniques, they stack genes for resistance to a wide variety of evolving rusts. Similarly, he says, "It will be possible to deploy different B.t. genes or stack genes and thereby stay ahead of the ever-evolving pest populations." Given their concerns about the spread of transgenes, you might think opponents of biotech would welcome innovations designed to keep transgenes confined. Yet opponents became apoplectic when Delta Pine Land Co. and the U.S. Department of Agriculture announced the development of the Technology Protection System, a complex of three genes that makes seeds sterile by interfering with the development of plant embryos. TPS also gives biotech developers a way to protect their intellectual property: since farmers couldn't save seeds for replanting, they would have to buy new seeds each year.

Because high-yielding hybrid seeds don't "breed true"—that is, the progeny of the crossbred hybrids will exhibit an unpredictable and undesirable mixture of the parental stocks' characteristics—corn growers in the United

States and western Europe have been buying seed annually for decades. Thus TPS seeds wouldn't cause a big change in the way many American and European farmers do business. If farmers didn't want the advantages offered in the enhanced crops protected by TPS, they would be free to buy seeds without TPS. Similarly, seed companies could offer seeds with transgenic traits that would be expressed only in the presence of chemical activators that farmers could choose to buy if they thought they were worth the extra money. Ultimately, the market would decide whether those innovations were valuable.

If anti-biotech activists really are concerned about gene flow, they should welcome such technologies. The pollen from crop plants incorporating TPS would create sterile seeds in any weed with which the crop plant happened to crossbreed, so that genes for traits such as herbicide resistance or drought tolerance couldn't be passed on. That point escapes some opponents of biotech. "The possibility that TPS may spread to surrounding food crops or to the natural environment is a serious one," writes Indian anti-biotech activist Vandana Shiva in her recent book *Stolen Harvest*. "The gradual spread of sterility in seeding plants would result in a global catastrophe that could eventually wipe out higher life forms, including humans, from the planet." That dire scenario is not just implausible but biologically impossible.

TPS is a gene technology that causes sterility; *that means, by definition, that it can't spread.* Despite the clear advantages that TPS offers in preventing the gene flow that activists claim to be worried about, the Rural Advancement Foundation International, now the ETC Group, quickly demonized TPS by dubbing it 'Terminator Technology.' RAFI warned "If the Terminator Technology is widely utilized, it will give the multinational seed and agrochemical industry an unprecedented and extremely dangerous capacity to control the world's food supply." Responding to activist protests, Monsanto, which had acquired the technology when it bought Delta Pine Land Co., declared that it would not develop TPS.

Even so, researchers have developed another clever technique to prevent transgenes from getting into weeds through crossbreeding. Chloroplasts (the little factories in plant cells that use sunlight to produce energy) have their own small sets of genes. Researchers can introduce the desired genes into chloroplasts instead of into cell nuclei where the majority of a plant's genes reside. The trick is that the pollen of most crop plants doesn't have chloroplasts; therefore it is impossible for a transgene confined to chloroplasts to be transferred through crossbreeding.

## Public opinion vs. sound science

To date, the American public and policymakers have not generally succumbed to the scares and bogus concerns being peddled by anti-biotech activists. Europe, however, is another matter entirely. A recent poll in the United Kingdom found that 51 percent of British consumers would avoid eating genetically enhanced foods, while 40 percent would not. However, 76 percent of respondents favoured labeling biotech foods, while only 6 percent agreed with the U.S. view that such foods should not be labeled. Since it is widely agreed by scientific experts around the world and U.S. regulatory authorities that food produced using biotech crops is safe, why are European regulators, who know that the technology is safe, trying to ban it or stigmatize it using labels that the public would likely misconstrue as warning labels? Concern about competition is certainly one often-unstated reason European governments have been so quick to oppose crop biotechnology. "EU countries, with their heavily subsidized farming, view foreign agribusinesses as a competitive threat," Frances Smith, director of Consumer Alert, has written. "With heavy subsidies and price supports, EU farmers see no need to improve productivity." In fact, biotech-boosted European agricultural productivity would be a fiscal disaster for the EU, since it would increase already astronomical subsidy payments to European farmers.

Currently, the EU's Common Agricultural Policy subsidy payments make up half of the EU's entire budget. Eighty percent of the EU subsidies go to just 20 percent of European farmers, generally those with the largest farms. EU agricultural policy is hostage to member-state concerns, much as U.S. farm policy is hostage to the demands of senators from sparsely populated farming states.

### Where does the trade battle stand?

The battle over biotech crops is now being joined in virtually all of the institutions that govern the world's food trade system, including the World Trade Organization, the new Biosafety Protocol, and the Codex Alimentarius Commission. European resistance to genetically enhanced crops is generally traced to the concerns about food safety that erupted with the outbreak of mad cow disease in Britain and food contamination problems in Belgium in the 1990s. But there is a longer history to the EU's hostility to biotech.

Starting in 1990, EU regulators used specious health concerns to fight against the importation of American beef and milk produced using biotech bovine growth hormone. The EU suffered a string of losses in international arbitration, first under the General Agreement on Tariffs and Trade and then in the WTO, which finally ruled in 1999 that the United States could impose punitive duties on more than $100 million in European exports in retaliation.

## Precautionary principle paralysis

The EU is justifying its ban of and import restrictions on biotech crops on the basis of the 'precautionary principle.' Under that principle, regulators do not need to show scientifically that a biotech crop is unsafe before banning it; they need only assert that it has not been proved harmless.

"They want to err on the side of caution not only when the evidence is not conclusive but when no evidence exists that would indicate harm is possible," observes Smith. The preçautionary principle is best summed up as 'regulate first, ask questions later.' The strictest interpretations of the precautionary principle jettison entirely the notion of tradeoffs, requiring that any new technology never cause any harm to the environment or human health. Of course, accurately predicting in advance the benefits and harms that a technology may one day produce is an impossible task. This inherent uncertainty means that opponents of a new technology can always stall its introduction by endlessly demanding that more research be done to rule out even their most farfetched fears.

As researchers Soren Holm and John Harris explained in *Nature*: As a principle of rational choice, the PP will leave us paralyzed. In the case of genetically modified (GM) plants, for example, the greatest uncertainty about their possible harmfulness existed before anybody had yet produced one. The PP would have instructed us not to proceed any further, and the data to show whether there are real risks would never have been produced.

The same is true for every subsequent step in the process of producing GM plants. The PP will tell us not to proceed, because there is some threat of harm that cannot be conclusively ruled out, based on the evidence from the preceding step. The PP will block the development of any technology if there is the slightest theoretical possibility of harm. So it cannot be a valid rule for rational decisions. In other words, the only way to protect completely against unknown risks is never to do anything for the first time. The precautionary principle certainly is irrational in scientific terms, but it is, unfortunately, all too rational in terms of satisfying the political needs of regulators. Under the WTO, the Sanitary and Phytosanitary Agreement allows countries to set their own health and environmental standards. But in Article 2.2, the SPS says regulations must be based on scientific principles and that they should not be maintained without sufficient scientific evidence. Therefore, it would seem that the SPS requirement that regulations be justified by scientific assessments would rule out the precautionary principle.

Similarly, under the WTO, the Technical Barriers to Trade agreement requires that countries avoid unnecessary obstacles to trade in adopting regulations aimed at protecting human health and safety or the environment.

Regulators may set general standards but not specify how a product should be made. For example, a country could adopt a safety regulation that says that a door must resist fire for 30 minutes but not one that says the door must be made of steel. Clearly, a rule requiring that food be safe is acceptable, but one banning foods made from genetically modified crops is not, since all relevant scientific authorities agree that all approved genetically modified crops are healthy and safe for human consumption.

It is vital that the U.S. negotiators resist European efforts to undermine WTO agreements by reinterpreting them to include consumer desire for more information as a legitimate goal under the TBT.

## Protectionist labels

In any case, the EU is trying to make an end run around the relatively clear standards set out by the WTO through two other international forums, the Cartagena Biosafety Protocol and the Codex Alimentarius Commission. The Biosafety Protocol was drafted under the Convention on Biological Diversity (never ratified by the United States) and completed in 2000. The protocol, largely negotiated by environment ministers rather than trade ministers, focuses almost entirely on international trade in 'living genetically modified organisms'(LMOs). Specifically, that means trade in genetically enhanced crops and livestock.

The Biosafety Protocol specifically incorporates the precautionary principle in its preamble and in Articles 10 and 11 as justification for signatories to limit the importation of LMOs such as grains and livestock. Article 18 of the protocol also allows importing countries to require that shipments containing LMOs, say genetically enhanced corn or soybeans; be labeled may contain LMOs. Furthermore, the Biosafety Protocol ambiguously states that it should not be interpreted as implying a change in the rights and obligations of a party under any existing international agreements, but also that it is not "subordinate . . . to other international agreements."

The Biosafety Protocol requires that all shipments of biotech crops, including grains and fresh foods, carry a label saying that they may contain living modified organisms. This international labeling requirement is clearly intended to force the segregation of conventional and biotech crops. The protocol was hailed by Greenpeace's Benedikt Haerlin as a historic step towards protecting the environment and consumers from the dangers of genetic engineering. Shortly after the Biosafety Protocol negotiations were completed in 2000, the European Commission issued a Communication from the Commission on the Precautionary Principle, explaining how the EU would incorporate the principle in its regulatory systems.

The communication explicitly noted, "The concept of risk in the SPS leaves leeway for interpretation of what could be used as a basis for a precautionary approach." And further noted that international standards recognized under the SPS were being negotiated at the Codex Alimentarius Commission. Using the Biosafety Protocol and the Codex Commission's interpretation of the precautionary principle, in July 2001 the European Commission issued a set of draft regulations regarding biotech crops. Those regulations, which will come into effect in October 2002, impose 'traceability' and labeling requirements on all foods made using biotech crops, including imports. Traceability means that farmers and the food industry must create, retain, and transmit information about the origin of foods made using genetically enhanced crops at each stage of production and distribution (from dirt to fork). Industry must create systems that identify to whom and from whom products using biotech crops are made available. That information must be transmitted throughout the commercial chain and must be retained for five years. In addition, all foods produced using ingredients derived from biotech crops and live- stock, irrespective of whether they actually contain genetically modified DNA or proteins in the final product, must bear the following label: "This product contains genetically modified organisms." Even corn syrup and soybean oil, which contain no detectable levels of DNA or biotech-derived proteins, will have to be labeled, in this case, erroneously, as containing genetically modified organisms. Similar requirements are proposed for feed grains that human beings will not eat.

The Codex Alimentarius Commission is an intergovernmental body created in 1962 to set food standards under the auspices of the UN's Food and Agriculture Organization and the World Health Organization. In 1995 the SPS agreement conferred on the Codex Commission the responsibility for setting international food safety standards that would be recognized by the WTO. EU negotiators are well on their way to persuading the Codex Commission to adopt standards that would require that foods that have genetically modified crops as ingredients carry mandatory labels and be able to be traced. U.S. negotiators from the FDA and the USDA have already given away the store by conceding to EU demands in the Codex Ad-Hoc Intergovernmental Task Force on Foods Derived from Biotechnology's most recent meeting in Yokohama, Japan, in March.

Paragraph 19 of the Draft Principles for the Risk Analysis of Foods Derived from Modern Biotechnology states, "Risk management measures may include, as appropriate, food labeling, conditions for marketing approvals and post-market monitoring." The EU's long-sought traceability provision is incorporated in Paragraphs 20 and 21 of the draft principles. Under the guise of risk management procedures, those articles permit regulators to set up mechanisms for the tracing of products for the purpose of facilitating withdrawal from the market when a risk to human health has been identified or to support post-market monitoring in circumstances as indicated in paragraph

20." These draft principles are being submitted for consideration of the commission in July 2003.

Furthermore, the EU negotiators have managed to incorporate the precautionary principle in the codex deliberations by persuading the Codex Committee on General Principles to forward Proposed Draft Working Principles for Risk Analysis in the Framework of the Codex Alimentarius to the commission's executive committee for adoption as draft principles. This risk analysis draft specifically acknowledges, "Precaution is an inherent element of risk."

Environmental groups such as the International Union for the Conservation of Nature and the International Centre for Trade and Sustainable Development are already hailing the draft principles as a European victory in the attempt to limit trade in genetically enhanced crops. "Some observers believe that the agreement reached at the Codex Commission meeting might mark a breakthrough in international negotiations on the use of traceability systems and at least partially vindicates the EU's insistence on introducing a labelling and traceability system for GM foods," notes ICTSD's *Bridges* report on trade. Clearly, U.S. trade interests are not being well served by allowing the FDA and the USDA to take the lead in the codex negotiations.

Despite the fact that the European Commissioner for Health and Consumer Protection, David Byrne, admitted last October that there is an irrational fear of GM food in the EU, he justified these proposed regulations on consumer choice and protection grounds. Indeed, even if no hazards from genetically improved crops have been demonstrated, don't consumers have a right to know what they're eating? This seductive appeal to consumer rights has been a very effective public relations gambit for anti-biotech activists and European bureaucrats eager to expand their jurisdictions. If there's nothing wrong with biotech products, they ask, why shouldn't seed companies, farmers, and food manufacturers agree to label them?

The activists are being more than a bit disingenuous here. Their scare tactics, including the use of ominous words such as *Frankenfoods*, have creat ed a climate in which many consumers would interpret labels on biotech products to mean that they were somehow more dangerous or less healthy than oldstyle foods. Opponents of biotech hope labels will drive frightened consumers away from genetically modified foods and thus doom them. Then the activists could sit back and smugly declare that biotech products had failed the market test.

## An organic alternative to GMO labels

In the United States, the biotech labeling campaign is a red herring, because the USDA, at the insistence of organic farmers, has issued some 554 pages of regulations outlining which foods qualify as 'organic.'65 Among other things, the definition requires that organic foods not be produced using genetically modified crops. Thus U.S. consumers who want to avoid biotech products need only look for the 'organic' label. Furthermore, there is no reason why conventional growers who believe they can sell more by avoiding genetically enhanced crops should not label their products accordingly, so long as they do not imply any health claims. The FDA has begun to solicit public comments on ways to label foods that are not genetically enhanced without implying that they are superior to biotech foods. The European Union could adopt this approach instead of imposing new regulations on genetically enhanced crops and foods.

In any case, labeling nonbiotech foods as such will not satisfy the activists whose goal is to force farmers, grain companies, and food manufacturers to segregate biotech crops from conventional crops. Such segregation would require a great deal of duplication in infrastructure, including separate grain silos, rail cars, ships, and production lines at factories and mills. It has been estimated that constructing the parallel infrastructure needed to comply with these regulations could cost the American farm sector as much $4 billion. The StarLink corn problem is just a small taste of how costly and troublesome segregating conventional from biotech crops would be. Some analysts estimate that segregation would add 12 percent to grain prices without any increase in safety.

Activists are also clearly hoping that mandatory crop segregation will lead to novel legal nightmares: If a soya bean shipment is inadvertently 'contaminated' with biotech soya beans, who is liable? If biotech corn pollen falls on an organic cornfield, can the organic farmer sue the biotech farmer? Consider the Catch-22 situation that organic farmers have created for themselves. As the editors of *Nature Biotechnology*note: "Organic certification is a form of self-regulation imposed, in essence, by organic farmers on organic farmers.

The rules have been established so that all organic farmers play on a level field. 'No GM' is one of these rules. Having established themselves as rule-makers, law enforcement agencies, and juries, organic certification bodies are now endeavouring to obtain judgments from legislative bodies that had no part in establishing the rules in the first place." Even worse than the proposed EU regulations are model biosafety laws proposed by the activist group Third World Network. Under the model legislation, "The absence of scientific evidence or certainty does not preclude the decision makers from denying approval of the introduction of the GMO or derived products." Worse, under the model regulations, "Any adverse socio-economic effects must also be considered." In April 2001 the EC issued a directive covering genetically enhanced foods, which also directs regulators to take into account the socio economic effects of introducing biotech crops and foods. If provisions like these are adopted, they could give traditional producers a veto over innovative competitors, the moral equivalent of letting candle makers prevent the introduction of electric lighting.

The Office of the U.S. Trade Representative has suggested several times that it might ask the WTO to adjudicate these issues, but so far it has not taken any concrete steps to do so despite the obvious dangers posed to U.S. agricultural exports.

## Farm trade and scientific standards at risk

The brewing U.S.-EU trade war over biotech crops could imperil the whole WTO system of international trade, especially if socio economic considerations are incorporated into any trade rule negotiations in the guise of implementing the precautionary principle. As the U.S.-EU dispute over importing U.S. beef produced using growth hormone indicates, the EU seems willing to accept the imposition of punitive duties against its exports rather than comply with WTO rulings.

Time is of the essence. U.S. trade negotiators who are relying on the WTO and SPS provisions to challenge European efforts to limit trade in genetically enhanced crops are about to be blindsided by European negotiators' efforts to subvert the SPS Agreement's scientific standards through the codex negotiations. How? The SPS Agreement recognizes the Codex Commission as the international organization responsible for setting standards related to food safety. According to the SPS Agreement, WTO members 'shall base' their measures related to human and plant health on codex standards, guidelines, or recommendations. Codex measures shall be deemed to be necessary to protect human, animal or plant life or health, and presumed to be consistent with the relevant provisions of the SPS Agreement.

As noted above, the Codex Commission will meet in Rome in

July 2003 to consider approving the draft principles governing foods derived from modern biotechnology. If the commission at that time accepts the draft standards, that would provide the EU a rationale for imposing labeling and tracing requirements of biotech foods under the guise of meeting international health and safety standards under the SPS Agreement. In that case, the United States would likely lose any future WTO challenge to EU labeling and traceability requirements imposed on imports of U.S. biotech crops. That must not be allowed to happen.

Fortunately, the U.S. trade negotiators can stop the codex process in its tracks. All codex standards must be agreed to by consensus of all the parties. U.S. negotiators must simply call a halt at the Rome meeting to inclusion of the precautionary principle, biotech labeling, and traceability requirements. Any language incorporating notions of the precautionary principle must be ripped out of codex principles root and branch. U.S. negotiators must make it clear that only science-based risk standards will be acceptable in protecting human health and food safety.

It is vital that the Office of the U.S. Trade Representative recognize that the Europeans are clearly no longer treating the Codex Commission as a forum for setting international food safety and health standards. They are treating it as an alternative forum for international trade negotiations. This means that the USTR must take the lead over the FDA and the USDA in negotiations of the Codex Commission beginning at the July meeting. The EU has made it clear that it intends to use the Biosafety Protocol's recognition of the precautionary principle as a justification for imposing labeling and traceability requirements on biotech crop imports. The Biosafety Protocol will come into effect internationally only if it is ratified by 50 nations. So far it has been ratified by 11 nations. This gives U.S. negotiators an opportunity to make concerted efforts to persuade developing nations not to ratify the Biosafety Protocol and to approve biotech crops for domestic production and consumption, thus isolating the EU. This effort could perhaps be coordinated by the U.S.

Department of State's Bureau of Economicand Business Affairs and the USTR. By stopping the adoption of codex draft principles and making a concerted effort to prevent the Biosafety Protocol from coming into effect, the United States will make it clear to EU officials that their proposed biotech regulations will be challenged. In the face of this challenge, EU officials may be persuaded to rethink and revise their proposed regulations before putting them into effect this October.

One final possibility is that the USTR could bring the EU labeling and traceability regulations for adjudication by the WTO. However, if the Codex Commission adopts the draft principles discussed above, the United States could well lose at the WTO. Even if the USTR does derail the codex draft principles and does win at the WTO, such a victory could turn into a public relations disaster as European governments tell their citizens that American corporations are forcing genetically enhanced foods down their throats. Still, the USTR must be willing to take this step if all else fails.

## Conclusion

It is essential to preserve and insist upon standards based on scientific risk assessment in order to maintain and expand a freer international trading system. Jettisoning scientific risk assessment and replacing it with a precautionary approach will open the entire trading system to interruptions based on arbitrary justifications. Capricious labeling requirements will also proliferate. Such labels are unjustifiabiy stigmatizing and costly and offer no consumer health or safety benefits. Only objective scientific standards should be used because regulations that are based on 'societal values' alone can never be agreed upon internationally and will restrict trade without protecting public safety.

Consequently, all U.S. negotiators involved with trade in biotech crops must make it unalterable U.S. policy to oppose the application of the precautionary principle and instead insist on scientifically based risk standards in all international trade for. Among other things, this means that all U.S. trade negotiators in whatever for must insist that no labeling or traceability requirements be imposed on biotech food products that are substantially equivalent to nonbiotech crops.

**CHAPTER-9**

# The Role of Biotechnology in The Socio-Economic

## Development

Biotechnology is any technique which involves the application of biological organisms or their components, systems or processes to manufacturing and service industries make or modify products, to improve plants or animals or to develop micro-organisms for special uses. Some people use the term biotechnology to refer to the tools of genetic engineering that have been developed since 1973. But biology, technology, and human-directed genetic change have been a part of agriculture since the beginning of cultivated crops some 10,000 years ago. Biotechnology has, in a general sense, been used as a tool for food production since the first breeders decided to selectively plant or breed only the best kinds of corn or cows. Technology is a tool we use to achieve a goal, such as improved food quality.

Scientific advances through the years have relied on the development of new tools to improve socio-economy such as health care, agricultural production, and environmental protection. Individuals, consumers, policymakers, and scientists must ultimately decide if the benefits of biotechnology are greater than the risks associated with this new approach. The technology tools used in biology have changed rapidly since scientists moved the first specific gene from one organism to another in 1973. This new era began in 1953 when scientists James Watson and Francis Crick determined the structure of DNA. DNA is the chemical language that determines the features and characteristics of all living organisms: plants, animals, and microorganisms. Once scientists understood how DNA was put together, they could determine which parts of the DNA (genes) are responsible for certain traits.

Genes determine traits by controlling the production of proteins, including enzymes. Proteins and enzymes are used by all living organisms to grow, metabolize energy, and become what their genetic code dictates. Each cell of an organism contains the entire genetic code needed to create the organism. The interaction of genetic makeup and environmental factors shapes the nature of all living things. Whcn pcoplc cat a 'hcalthy' diet, they are controlling environmental factors that will, within the limits of their genetic makeup, decrease their risk of developing a disease.

In Europe, a vast diversity of high quality foods provide the carbohydrates, fats, proteins, minerals and vitamins needed in the everyday diet of consumers. At the heart of food production is biotechnology. One aspect of biotechnology which has been used for centuries is the selective breeding of crop plants and farm animals to produce improved food. Another is fermentation, in use for millennia to produce fermented foods like cheese, bread, beer, sauerkraut and sausages.

Gene technology includes any biotechnological technique for the controlled modification or transfer of genes from one organism to another to give a desired characteristic. The first use of gene technology two decades ago opened up the potential for many additional advances in both selective breeding and fermentation. Each specific step forward might be relatively small, but together they could add up to further improvements in the nutritional quality, appearance, flavour, convenience, cost and safety of foods which are an integral part of socio-economic advancement and national development..

This overview seeks to provide science-based information about discoveries in biotechnology as it affects mankind and is designed to help us understand and assess the risks and benefits of biotechnology. It also provides information about biotechnology with examples of how these new tools of biology and agriculture are used in food production as part of socio-economic advancement and national development. It includes a perspective showing how biotechnology fits into the history and future of science and food for mankind. Its purpose is to educate people about biotechnology and its role in the socio-economic advancement and development of a nation so that they can make informed choices.

**Biotechnology applications in food processing-** Biotechnology includes a wide range of diverse technologies and they may be applied in each of the different food and agriculture sectors. It includes technologies such as gene modification (manipulation) and transfer, the use of molecular markers, development of recombinant vaccines and DNA-based methods of disease characterization/ diagnosis, *in vitro* vegetative propagation of plants, embryo transfer and other reproductive technologies in animals or triploidization in fish. It also includes a range Okonko et al. 2355 of technologies used to process the raw food materials produced by the crop, fishery and livestock sectors. This is the area that has received relatively little attention from the media, but which is very important for food security in many developing countries. Biotechnology in the food processing sector targets the selection and improvement of microorganisms with the objectives of improving process control, yields and efficiency as well as the quality, safety and consistency of bio-processed products.

Microorganisms or microbes are generic terms for the group of living organisms which are microscopic in size, and include bacteria, yeasts and moulds. Fermentation is the process of bioconversion of organic substances by microorganisms and/or enzymes (complex proteins) of microbial, plant or animal origin. It is one of the oldest forms of food preservation which is applied globally. Indigenous fermented foods such as bread, cheese and wine, have been prepared and consumed for thousands of years and are strongly linked to culture and tradition, especially in rural households and village communities. It is estimated that fermented foods contribute to about onethird of the diet worldwide. During fermentation processes, microbial growth and metabolism (the biochemical processes whereby complex substances and food are broken down into simple substances) result in the production of a diversity of metabolites (products of the metabolism of these complex substances). These metabolites include enzymes which are capable of breaking down carbohydrates, proteins and lipids present within the substrate and/or fermentation medium; vitamins; antimicrobial compounds (e.g. bacteriocins and lysozyme), texture-forming agents (e.g. xanthan gum), amino acids, glutamic acids organic acids (e.g. citric acid, lactic acid) and flavour compounds (e.g. esters and aldehydes). Many of these microbial metabolites (e.g. flavour compounds, amino acids, organic acids, enzymes, xanthan gums, alcohol, etc.) are produced at the industrial level in both developed and developing countries for use in food processing applications. A considerable volume of current research both in academia and industry targets the application of microbial biotechnology to improve the production, quality and yields of these metabolites.

Some researches have been carried out on the production of fermented condiments, iru from African locust bean melon seed fermented ogiri and soya bean produced daddawa. Some of the most important food condiments are ogiri, which is produced from melon seeds (*Citrullus vulgaris)*, iru or dawadawa produced from locust bean (*Parkia biglobosa)* Odunfa, ugba produced from oil bean seeds (*Pentaclethra macrophylla*), ogiri-Igbo is produced from castor oil seeds (*Ricinus* 2356 Afr. J. Biotechnol. *communis*) Odunfa Dawadawa from soya beans. Iru from Locust beans Owoh from cotton seeds (*Gossypium hirsutum*), Okpehe from mesquite (*Prosopis africana*). Ogiri which is made from (*Parkia filicoidea)*, Melon seeds (*Citrullus vulgaris)* is also produced from Africa oil bean (*Pentaclethra macrophylla*) and from Asia oil seeds (*Albizia saman*). They are generally proceesed to yield condiments (Table 1). These Africa fermented condiments requires food processing technologies that will meet the requirement/challenges of human needs. Although, fermentation rightly meets these needs of mankind. Fermentation improves markedly the digestion, nutritive value, and flavour of the raw materials used.

Food condiments in Nigeria and many other countries of West and Central Africa are popular strong-smelling fermented food culinary products that give pleasant aroma to soups, sauces and other prepared dishes. They also have great potential as key protein, fatty acid are good sources of gross energy. Therefore, condiments are basic ingredients for food supplementation and their socio-economic importance cannot be over emphasized in many countries especially in Africa and Asia (India) where protein calorie malnutrition is a major problem.

Fermented foods have a long history in Africa. However, the absence of writing culture in most African countries makes their origin difficult to trace. Perhaps the most documented of the fermented food is sour-milk, followed in historical importance by various alcohol drinks which played important part on various solemn occasions and over thirty different fermented foods have been recorded.

Fermentation is globally applied in the preservation of a range of raw agricultural materials (cereals, roots, tubers, fruit and vegetables, milk, meat, fish etc.). Commercially produced fermented foods which are marketed globally include dairy products (cheese, yogurt, and fermented milks), sausages and soy sauce. Certain microorganisms associated with fermented foods, in particular are strains of the *Lactobacillus* species, which are probiotics i.e. used as live microbial dietary supplements or food ingredients that have a beneficial effect on the host by influencing the composition and/or metabolic activity of the flora of the gastrointestinal tract. Probiotic bacterial strains are also produced and commercially marketed in many developed countries. In developing countries, fermented foods are produced primarily at the household and village level, where they find wide consumer acceptance.

Food fermentations contribute substantially to food safety and food security, particularly in the rural areas of many developing countries. Traditional Fermentation processes used in the production of these foods are uncontrolled and are dependent on microorganism.

**Table 1.** Microbial fermentation is essential to the production of these fermented foods

| S/No. | Fermented food | S/No. | Fermented food |
|---|---|---|---|
| 1 | Aisa | 17 | Kefir |
| 2 | Beer | 18 | Miso |
| 3 | Bologna | 19 | Pap |
| 4 | buttermilk | 20 | Pickles |
| 5 | Cheeses | 21 | Salami |
| 6 | Cider | 22 | Sauerkraut |
| 7 | Cocoa | 23 | Sour cream |
| 8 | Coffee | 24 | Soy sauce |
| 9 | Cottage cheese | 25 | Tamari |
| 10 | Distilled liquors | 26 | Tea |
| 11 | Dawadawa/daddawa | 27 | Tempeh |
| 12 | Iru | 28 | Tofu |
| 13 | Ogiri | 29 | Ugba |
| 14 | Okpehe | 30 | Vinegar |
| 15 | Olives | 31 | Wine |
| 16 | Owoh | 32 | Yogurt |

from the environment or the fermentation substrate for initiation of the fermentation processes. Such processes, therefore, result in products of low yield and variable quality. Microorganisms and metabolic pathways associated with the production of fermented foods are the subject of considerable research, targeting strain isolation and identification; improvement of the efficiency of fermentation processes and the quality, safety and consistency of fermented foods. Much of this research incorporates the use of genetic technologies for strain development and improvement, and for diagnostic studies.

While microorganisms are beneficial in most fermentation processes, some may pose the risk of food contamination and can cause food-borne illness. Diagnostic methodologies which integrate the use of molecular genetic techniques enhance the speed and sensitivity of microbial testing and are increasingly being applied in developing countries. The application of biotechnology to the processing of food (including beverages) produced from agriculture involves biotechnological tools and options that are applicable to the study and improvement of microorganisms which offer potential for improving the quality, safety and consistency of fermented foods; improving efficiency in the production of fermented foods, food ingredients, food additives and food processing aids (enzymes); diversifying thc outputs of fermentation processes and, finally, improving diagnostic and identification systems applicable to foods.

## Advances in pecific areas of food processing

## Improved food ingredients

Necessary changes to the key food ingredients, starches and oils, are usually made by processing. Biotechnology opens up the possibility of altering crop plants to produce exactly the type of ingredients needed:

*Starches*— Plant breeders have introduced a bacterial gene into potato plants which increases the proportion of starch in the tubers whilst reducing their water content. This means that the potatoes absorb less fat during frying, giving low-fat chips. Sweeter potatoes have also been produced which have higher sucrose content than traditional varieties.

*Oils*—Both rapeseed and sunflower are being altered to produce more stable and nutritious oils, which contain linoleic acid instead of linolenic acid and have lower saturated fat content. Rapeseed has also been modified to produce a high-temperature frying oil low in saturated fat.

## Product quality

Biotechnology has been employed to change the characteristics of the raw material inputs so that they are more attractive to consumers and more amendable to processing. Biotechnology researchers are increasing the shelf life of fresh fruits and vegetables; improving the crispness of carrots, peppers and celery; creating seedless varieties of grapes and melons; extending the seasonal geographic availability of tomatoes, strawberries and raspberries; improving the flavour of tomatoes, lettuce, peppers, peas and potatoes; and creating caffeine-free coffee and tea. Japanese scientists have now identified the enzyme that produces the chemical that makes us cry when we slice an onion. Knowing the identity of the enzyme is the first step in finding a way to block the gene to create 'tearless' onions.

Much of the work on improving how well crops endure food processing involves changing the ratio of water to starch. Potatoes with higher starch content are healthier because they absorb less oil when they are fried. Another important benefit is that starchier potatoer require less energy to process and therefore, cost less to handle. Many tomato processors now use tomatoes derived from a biotechnology technique, somaclonal variant selection. The new tomatoes, used in soup, ketchup and tomato paste, contain 30 percent less water and are processed with greater efficiency. A 1.2% increase in the solid content is worth $35 million to the U.S. processed-tomato industry.

Another food processing sector that will benefit economically from better quality raw materials is the dairy products industry. Scientists in New Zealand have now used biotechnology to increase the amount of the protein casein, which is essential to cheese making, in milk by 13 percent.

## Better raw materials

In improving raw food materials, many plant breeding programmes have been directed towards boosting yield or allowing more environmentally compatible agriculture by increasing the resistance of crops to viruses, pests or herbicides. Increasing yield has clear benefits in helping to feed the world's ever-increasing population and could provide cheaper food. Plants which are resistant to attack by insect pests and diseases would need fewer pesticide applications; resistant crops such as maize, tomatoes and potatoes are already being developed. Crops have also been produced with tolerance to modern, more environmentally compatible herbicides, with the aim of achieving optimal weed control with reduced levels of herbicide. Today, there is increasing interest in improving the nutritional value, flavour and texture of raw materials. This could help encourage greater fruit and vegetable consumption in line with government guidelines on healthy nutrition.

## Safety of the raw materials

The most significant food-safety issue food producers' face is microbial contamination, which can occur at any point from farm to table. Any biotechnology product that decreases microbes found on animal products and crop plants will significantly improve the safety of raw materials entering the food supply. Improved food safety through decreased microbial contamination begins on the farm. Transgenic disease-resistant and insect-resistant crops have less microbial contamination.

New biotechnology diagnostics detect microbial diseases earlier and more accurately, so farmers can identify and remove diseased plants and animals before others become contaminated.

Biotechnology is improving the safety of raw materials by helping food scientists discover the exact identity of the allergenic protein in foods such as peanuts, soya beans and milk, so they can then remove them. Although 95 percent of food allergies can be traced to a group of eight foods, in most cases we do not know which of the thousands of proteins in a food triggered the reaction. With biotechnology techniques, great progress are being made in identifying these allergens. More importantly, scientists have succeeded in using biotechnology to block or remove allergenicity genes in peanuts, soya beans and shrimping. Finally, biotechnology is helping us improve the safety of raw agricultural products by decreasing the amount of natural plant toxins found in foods such as potato and cassava.

## Enhanced food safety

In addition to the many ways biotechnology is helping us enhance the safety of the food supply; biotechnology is providing us with many tools to detect microorganisms and the toxins they produce. Monoclonal antibody tests, biosensors, polymerase chain reaction (PCR) methods and DNA probes are being developed and will be used to determine the presence of harmful bacteria that cause food poisoning and food spoilage, such as *Listeria* and *Clostridium botulinum.* In addition, *E. coli* 0157:H7, the strain of *E. coli* responsible for several deaths in recent years, can now be distinguished from the many other harmless *E. coli* strains. These tests are portable, quicker and more sensitive to low levels of microbial contamination than previous tests because of the increased specificity of molecular technique. For example, the new diagnostic tests for *Salmonella* yield results in 36 h compared with the three or four days the older detection methods required. Biotechnology-based diagnostics have also been developed that they allow us to detect toxins, such as aflatoxin, produced by fungi and molds that grow on crops, and to determine whether food products have inadvertently been contaminated with peanuts, a potent allergen.

## Improving food fermentors

Because of the importance of fermented foods to so many cultures, scientists are conducting a lot of work to improve the microorganisms that carry out food fermentations. The bacterium responsible for many of our fermented dairy products, such as cheese and yogurt, is susceptible to infection by a virus that causes substantial economic losses to the food industry. Through recombinant technology, researchers have made some strains of this bacterium and other important fermentors resistant to viral infection. It has been known for years that some bacteria used in food fermentation produce compounds that kill other contaminating bacteria, which cause food poisoning and food spoilage. Using biotechnology we are equipping many of our microbial fermentors with this self-defense mechanism to decrease microbial contamination of fermented foods.

## Advances in processing aids and additives

Microorganisms have been essential to the food industry not only for their importance as fermentors, but also because they are the source of many of the additives and processing aids used in food processing. Biotechnology advances will enhance their value to the food industry even further. Food additives are substances used to increase nutritional value, retard spoilage, change consistency and enhance flavour. The compounds food processors use as food additives are substances nature has provided and are usually of plant or microbial origin, such as xanthan gum and guar gum, which are produced by microbes. Many of the amino acid supplements, flavors, flavor enhancers and vitamins added to breakfast cereals are produced by microbial fermentation. Through biotechnology, food processors will be able to produce many compounds that could serve as food additives but that now are in scanty supply or that are found in microorganisms or plants difficult to maintain in fermentation systems. Food processors use plant starch as a thickener and fat substitute in low-fat products. Currently, the starch is extracted from plants and modified using chemicals or energy-consuming mechanical processes. Scientists are using biotechnology to change the starch in crop plants so that it no longer requires special handling before it can be used.

Traditional biotechnology has played a major role in producing fermented foods-where desirable changes are produced by the action of micro-organisms or enzymes; of which over 3,500 different types exist around the world. In Europe and North America, bread, yoghurt and cheese are perhaps most familiar. In Africa, foods made from fermented starch crops like yams and cassava are more important, whereas in Asia, products derived from fermented soya beans or fish predominate.

Fermentation can make the food more nutritious, tastier or easier to digest, and it can enhance food safety. It also helps to preserve food and to increase its shelf-life, reducing the need for additives. Genetically improved strains of microbes can make a major contribution to these desirable properties. For many years, a wide range of additives, processing aids and supplements have been obtained from microbial sources by fermentation. Products from modern biotechnology include vitamins, citric acid, natural colourings, flavourings, gums and enzymes. Gums used as low-calorie thickening agents and low-calorie sweeteners from natural ingredients are also produced using modern biotechnology.

Enzymes, the naturally-occurring catalysts responsible for literally all the biochemical processes of life, are used in applications such as bakery and cheese making to improve texture, appearance and nutritional value, and to generate desirable flavours and aromas.

A second area where biotechnology has advantages is in improving the processes by which food is produced. Now, it can be used to develop mild, highly specific processes using modified micro-organisms and purer, cheaper enzyme products. These can offer better productivity, cost-effectiveness and energy efficiency than existing processes. They can produce top-quality foods with a reduced need for additives such as flavourings, and can also reduce the environmental impact of food processing. Finally, this overview covers applications of biotechnology to processing of food as well as processing of non-food agricultural products (e.g. timber) and applying biotcchnology to microorganisms for environmental purposes (bioremediation, biofuels etc.). Researches are underway at present, aiming to allow the production of better food raw materials by crop plants. However, some processing steps remain essential to bridge the gap between currently-available raw materials and the desired end-product.

## Current status of biotechnology in food processing

Microorganisms are an integral part of the processing system during the production of fermented foods. Microbial cultures can be genetically improved using both traditional and molecular approaches, and improvement of bacteria, yeasts and moulds is the subject of much academic and industrial research. Traits which have been considered for commercial food applications in both developed and developing countries include sensory quality (flavour, aroma, visual appearance, texture and consistency), virus (bacteriophage) resistance in the case of dairy fermentations, and the ability to produce antimicrobial compounds (e.g. bacteriocins, hydrogen peroxide) for the inhibition of undesirable microorganisms. In many developing countries, the focus is on the degradation or inactivation of natural toxins (e.g. cyanogenic glucosides in cassava), mycotoxins (in cereal fermentations) and anti-nutritional factors.

## Traditional approaches

Traditional methods of genetic improvement such as classical mutagenesis and conjugation have been the basis of industrial starter culture development in bacteria (a culture used to start a food fermentation is known as a starter culture), while hybridisation has been used in the Okonko et al. 2359 improvement of yeast strains which are widely applied industrially in baking and brewing applications.

## Classical mutagenesis

This involves the production of mutants by the exposure of microbial strains to mutagenic chemicals or ultraviolet rays to induce changes in their genomes. Improved strains thus produced are selected on the basis of specific properties such as improved flavour-producing ability or resistance to bacterial viruses. Such mutants may, however, show undesirable secondary mutations which can influence the behaviour of cultures during fermentation.

## Conjugation

This is a natural process whereby genetic material is transferred among closely related microbial species as a result of physical contact between the donor and the recipient microorganism. Conjugational gene exchange allows both plasmid-localised and chromosomal gene transfer (a plasmid is a circular self-replicating nonchromosomal DNA molecule found in many bacteria, capable of transfer between bacterial cells of the same species, and occasionally of different species).

## Hybridization (i.e. sexual breeding or mating)

Sexual reproduction in yeasts, and thus genetic recombination, has led to improvements in yeasts. For example, crossing of haploid yeast strains with excellent gassing properties and with good drying properties could yield a novel strain with both good gassing and drying properties (FAO, 2000, 2004).

## Genetic modification

Recombinant DNA approaches have been used for genetic modification of bacterial, yeast and mould strains to promote expression of desirable genes, to hinder the expression of others, to alter specific genes or to inactivate genes so as to block specific pathways. The successful application of genetic modification for food bio-processing applications requires the development and use of food grade vectors, i.e. plasmids which do not contain antibiotic resistant genes as markers and which consist of DNA sequences from microorganisms which are generally recognized as safe (GRAS). Genetically Modified (GM) yeasts that are appropriate for brewing and baking applications have been approved for use (e.g. 2360 Afr. J. Biotechnol. approval was granted in the United Kingdom for use of a GM yeast (*Saccharomyces cerevisiae*) in beer production, containing a transferred gene from the closely related *Saccharomyces diastaticus*, allowing it to better utilise the carbohydrate present in conventional feedstocks. None of these GM yeasts are, however, used commercially.

## Genetic characterization

The genetic characterization of microbial strains through the use of molecular diagnostic techniques can contribute tremendously to the understanding of fermentation processes. Molecular diagnostics provide outstanding tools for the detection, identification and characterization of microbial strains for bio-processing applications and for the improvement of fermentation processes. The application of these and other related techniques, along with the development of molecular markers for bacterial strains, greatly facilitates understanding of the ecological interactions of microbial strains, their roles, succession, competition and prevalence in food fermentations and allows the correlation of these features to desirable quality attributes of the final product.

## Genomics

In recent years, the genome sequences of many foodrelated microorganisms have been completed (e.g. *S.cerevisiae*, commonly known as baker's or brewer's yeast, was the first eukaryote to have its genome sequenced - in 1996) and large numbers of microbial genome sequencing projects are also underway. Functional genomics, a relatively new area of research, aims to determine patterns of gene expression and interaction in the genome, based on the knowledge of extensive or complete genomic sequence of an organism. It can provide an understanding of how microorganisms respond to environmental influences at the genetic level (i.e. by expressing specific genes) in different situations or ecologies, and should 'therefore' allow adaptation of conditions to improve technological processes. For a range of microorganisms, it is now possible to observe the expression of many genes simultaneously, even those with unknown biological functions, as they are switched on and off during normal development or while an organism attempts to cope with pathogens or changing environmental conditions.

Cooper describes the use of DNA macroarrays to analyze expression of all 4,290 genes of the model bacterium *E. coli* after 20,000 generations of evolution in a glucose-limited medium. Functional genomics can, for example, shed light on common genetic mechanisms which enable microorganisms to use certain sugars during fermentation, as well as on genetic differences allowing some strains to perform better than others. It holds great potential for defining and modifying elusive metabolic mechanisms used by microorganisms. Moving from the gene to the protein level, it should also be mentioned that proteomics, an approach aiming to identify and characterize complete sets of protein, and protein-protein interactions in a given species, is also a very active area of research which offers potential for improving fermentation technologies.

## Biotechnology applications in production of food ingredients

Flavouring agents, organic acids, food additives and amino acids are all metabolites of microorganisms during fermentation processes. Microbial fermentation processes are 'therefore' commercially exploited for production of these food ingredients. Metabolic engineering, a new approach involving the targeted and purposeful manipulation of the metabolic pathways of an organism, is being widely researched to improve the quality and yields of these food ingredients. It typically involves alteration of cellular activities by the manipulation of the enzymatic, transport and regulatory functions of the cell using recombinant DNA and other genetic techniques. Understanding the metabolic pathways associated with these fermentation processes, and the ability to redirect metabolic pathways, can increase production of these metabolites and lead to production of novel metabolites and a diversified product base.

## Biotechnology applications in diagnostics/food testing

Many of the classical food microbiological methods used in the past were culture-based, with microorganisms grown on agar plates and detected through biochemical identification. These methods are often tedious, labour intensive and slow. Genetic based diagnostic and identification systems can greatly enhance the specificity, sensitivity and speed of microbial testing. Molecular typing methodologies, commonly involving the polymerase chain reaction (PCR), ribotyping (a method to determine homologies and differences between bacteria at the species or sub-species (strain) level, using restriction fragment length polymorphism (RFLP) analysis of ribosomal ribonucleic acids (rRNA genes) and pulsedfield gel electrophoresis (PFGE, a method of separating large DNA molecules that can be used for typing microbial strains), can be used to characterise and monitor the presence of spoilage flora (microbes causing food to become unfit for eating), normal flora and microflora in foods. Random amplified polymorphic DNA (RAPD) or amplified fragment length polymorphism (AFLP) molecular marker systems can also be used for the comparison of genetic differences between species, subspecies and strains, depending on the reaction conditions used. The use of combinations of these technologies and other genetic tests allows the characterization and identification of organisms at the genus, species, sub-species and even strain levels, thereby making it possible to pinpoint sources of food contamination, to trace microorganisms throughout the food chain or to identify the causal agents of food borne illnesses. Monoclonal and polyclonal antibodies can also be used for diagnostics, e.g. in enzyme-linked immunosorbent assay.

Microarrays are biosensors which consist of large numbers of parallel hybrid receptors (DNA, proteins, oligonucleotides).

Microarrays are also referred to as biochip, DNA chip, DNA microarray or gene arrays and offer unprecedented opportunities and approaches to diagnostic and detection methods. They can be used for the detection of pathogens, pesticides and toxins and offer considerable potential for facilitating process control, the control of fermentation processes and monitoring the quality and safety of raw materials.

## Biotechnology applications in the enzymes production

Enzymes are biological catalysts used to facilitate and speed up metabolic reactions in living organisms. They are proteins and require a specific substrate on which to work. Their catalyzing conditions are set within narrow limits, e.g. optimum temperature, pH conditions and oxygen concentration. Most enzymes are denatured at temperatures above 42°C. However, certain bacterial enzymes are tolerant to a broader temperature range. Enzymes are essential in the metabolism of all living organisms and are widely applied as processing aids in the food and beverage industry.

In the past, enzymes were isolated primarily from plant and animal sources, and thus a relatively limited number of enzymes were available to the food processor at a high cost. Today, bacteria and fungi are exploited and used for the commercial production of a diversity of enzymes (Table 2). Several strains of microorganisms have been selected or genetically modified to increase the efficiency with which they produce enzymes. In most cases, the modified genes are of microbial origin, although they may also come from different kingdoms. For example, the DNA coding for chymosin, an enzyme found in the stomach of calves, that causes milk to curdle during the production of cheese, has been successfully cloned into yeasts (*Kluyveromyces lactis*), bacteria (*E. coli*) and moulds (*Aspergillus niger* var. *awamori*).

**Table 2.** Marketed enzymes produced using gene technology.

| Principal enzyme activity | Application |
|---|---|
| Alpha-acetolactate decarboxylase | Brewing |
| Alpha-amylase | Baking, brewing, distilling, starch |
| Catalase | Mayonnaise |
| Chymosin | Cheese |
| Beta-glucanase | Brewing |
| Alphaglucanotransferase | Starch |
| Glucose isomerase | Starch |
| Glucose oxidase | Baking, egg mayonnaise |
| Hemicellulase | Baking |
| Lipase | Fats, oils |
| Maltogenic amylase | Baking, starch |
| Microbial rennet | Dairy |
| Phytase | Starch |
| Protease | Baking, brewing, diary, distilling, fish, meat, starch, vegetable |
| Pullulanase | Brewing, starch |
| Xylanase | Baking, starch |

Chymosin produced by these recombinant microorganisms is

currently commercially produced and is widely used in cheese manufacture.

The industrial production of enzymes from microorganisms involves culturing the microorganisms in huge tanks where enzymes are secreted into the fermentation medium as metabolites of microbial activity. Enzymes thus produced are extracted, purified and used as processing aids in the food industry and for other applications. Purified enzymes are cell free entities and do not contain any other macromolecules such as DNA. Genetic technologies have not only improved the efficiency with which enzymes can be produced, but they have increased their availability, reduced their cost and improved their quality. This has had the beneficial impact of increasing efficiency and streamlining processes which employ the use of enzymes as processing aids in the food industry.

In addition, through protein engineering, it is possible to generate novel enzymes with modified structures that confer novel desired properties, such as improved activity or thermostability or the ability to work on a new substrate or at a higher pH. Directed evolution is one of the main methods currently used for protein engineering. This technique involves creating large numbers of new enzyme variants by random genetic mutation and subsequently screening them to identify the improved variants. This process is carried out repeatedly, thus mimicking natural evolution processes.

## Biotechnology application in environment

Biotechnology application to microorganisms for environmental purposes includes bioremediation, biofuels, etc. Bioremediation is often successful and the most inexpensive method, it is only one of many techniques for dealing with hazardous wastes. This biological treatment is desirable because it is inexpensive, can be done at the site of pollution, and causes minimal physical disturbance to the surrounding area compared to other methods. The biological treatment of the contaminated soil and water is increasingly gaining popularity and acceptance though the technological advancement involved favours the industrialized countries due to limited access to these environmental technologies by developing countries which often lack the environmental regulatory framework for the application of biological treatment.

Though, environmental problems are one of the most important concerns for society, especially the wastes produced by industrial activity. The fermentation processes of foods, industrial production of dyes and textile material as well as domestic activities generate effluents with a high organic load. Most of the organic matter in waste has been removed by means of conventional anaerobic and/or aerobic biodegradation treatments. The use of specific treatments, physicochemical or biological or to remove the colour compounds is needed, as well as to optimize the operation conditions that are directly related to the characteristics of the treated waste.

Basic biological treatment or bioremediation is preferred over and above the other methods of treatment technologies such as physical, chemical, and thermal which are fast and controllable but requires high energy and cost prohibitive, and the techniques involved in biological treatment are usually cheap and do not need extensive training and controls and its basic principles are of universal validity.

Since the environmental issues are the same, in many areas of the world, especially the developing countries, there is a current drive for maximum and optimum utilization of all the possible byproducts from forest-based industries, of which, many new plant materials have already attracted attention.

According to Sutherland et al. Jahn and Hamid and Jahn the use of natural coagulants of plant origin to clarify turbid surface waters or employed in the treatment of muddy river water is not a new idea and waste minimization is a major concern of the whole concept.

## Biotechnology in agriculture

Applications of biotechnology to plants or animals have improved their food processing properties (e.g. development of the Flavr Savr tomato variety, genetically modified to reduce its ripening rate) and the production of proteins from genetically modified (GM) microorganisms to improve plant or animal production (e.g. production of bovine somatotropin (BST), a hormone increasing milk production in dairy cows, by GM bacteria).

## Advances from plant breeders to gene jockeys

Plant breeders have for many years used tools and techniques such as selective hybridization, grafting and cell isolation to improve crop quality and yield. And these early agricultural scientists made great advances, producing juicy ears of corn instead of hard-kerneled corn, which must be ground into flour, and present-day kiwi fruits rather than the hard berry from which they were developed. Scientists using the relatively new tools of biotechnology have been called 'gene jockeys' because of the great degree of speed and control with which they can change the inherited traits of plants, animals, and microorganisms. Today scientists can identify the gene(s) responsible for specific characteristics, such as disease resistance or nutrient composition, and insert them into another organism. What once took decades now takes years and can be accomplished with greater accuracy.

One of the most striking differences between traditional breeding and the genetic engineering approach is that the source of genetic material need not come from the same species. This allows scientists to exchange genetic information between bacteria, plants, and animals (including humans). These new techniques have prompted considerable debate on the ethical and moral aspects of this branch of science. All living organisms share the same genetic language. In fact, you probably share about half of your genetic information with a tomato plant. And the genetic information from that tomato plant can function in a corn plant. New techniques even allow scientists to decide in which part of the plant tissue a trait should be expressed, such as the pulp versus the skin of an apple.

When considering the risks associated with these new tools of food production, consumers need to understand how these tools differ from traditional agricultural methods. With traditional breeding methods, for example, increased levels of naturally occurring toxins may result from cross breeding designed to improve a crop. Breeders spend years 'back-crossing' to rid the new plant of the undesired feature while maintaining the benefits of the hybrid. There are also risks associated with the current standard use of chemicals to allow crops to tolerate insects, infections, and adverse weather conditions.

## Plant foods

When working with plant foods, scientists seek to improve foods for the benefit of consumers, producers, or the environment. Consumers may benefit from improved nutrition or food quality. Producers may be able to grow crops under adverse conditions, such as drought. Some genetically engineered plant foods require significantly fewer chemical applications during growth and 'therefore' have less environmental impact. Scientists use their current knowledge of plant biology to help them decide how to improve plant traits for foods. In the case of the slow-ripening Flavr Savr™ tomato introduced in 1994 by Calgene Inc., which was one of the first food plants produced using the tools of biotechnology, scientists knew that a type of protein called an enzyme causes tomatoes to soften as they ripen. When they isolated the gene responsible for the softening enzyme and inserted it *backwards* into the tomato's genetic code, the resulting tomato maintained good eating quality for a longer time than regular tomatoes. This technique allows better tasting tomatoes to be grown and shipped to distant markets.

In 1986, a herbicide-resistant soya bean was created using the tools of biotechnology. After several years of tests and studies, the Food and Drug Administration (FDA) and the U.S. Department of Agriculture (USDA) granted approval in 1994. The Environmental Protection Agency (EPA) granted approval in 1995, and the new soya beans were grown commercially in 1996. Given the widespread use of soya bean products as food ingredients, it has been estimated that most U.S. food consumers in the year 2000 have eaten foods produced through genetic engineering. In 1997, 18 crop applications were approved by the U.S. agencies responsible for regulating biotechnology. An estimated 35 percent of the 1999 U.S. corn and 55 percent of the soya bean crop were grown from genetically modified seeds.

**Animal foods**

The first FDA-approved application of biotechnology for production of food animals was to modify a microorganism to make a hormone needed for milk production in dairy cows. This genetically modified organism (GMO) is a bacterium that can produce large quantities of the hormone for injection into dairy cows. An estimated one-third of U.S. milk is produced using the GMO -produced hormone, which increases milk production by 10 to 25 percent. Another GMO is used to produce about 75 percent of U.S. cheese by providing a necessary enzyme formerly harvested from the stomach lining of cows. In addition to the use of GMOs in animal food production, biotechnology can be used to create transgenic animals. But developments of this biotechnology application may be slow due to the generally greater difficulties in animal genetic engineering and to the social and ethical concerns of consumers about the animal food applications of biotechnology.

Nevertheless, some genetically modified food animals are under consideration for approval and marketing. An example is a salmon that grows to a marketable size more rapidly than regular salmon. Most transgenic animal research is for medical applications, as in the case of the cloned sheep 'Dolly,' where scientists are investigating cystic fibrosis disease.

## Promising crop plants

*Improved nutritional value*—Crops in development include soya beans with higher protein content; potatoes with more nutritionally available starch and with improved amino acid content; pulses such as beans which have been altered to produce essential amino acids; crops which produce betacarotene, a precursor of vitamin A; and crop plants with a modified fatty acid profile. An example is a strain of oilseed rape which produces a special type of polyunsaturated fatty acid (the so-called w3-fatty acids). These have been linked to brain development and have potential in a range of speciality, clinical and infant foods.

*Better flavour*—Different types of peppers and melons with improved flavour are currently in field trials. Flavour can also be improved by enhancing the activity of plant enzymes which transform aroma precursors into flavouring compounds.

*Improved keeping properties*—There is improved keeping properties with the aim of making transport of fresh product easier, giving consumers access to nutritionally valuable whole foods and preventing decay, damage and loss of nutrients. Examples include the improved tomatoes now being sold in the US, and recently approved in the UK, which have been genetically altered to delay softening. Research is underway on making similar modifications to broccoli, celery, carrots, melon and raspberries. The shelf-life of some processed foods such as peanuts has also been improved by using raw materials with a modified fatty acid profile.

*Reduced levels of toxicants*—There is reduced level of toxicants thereby allowing a wider range of plants to be used as food crops, such as the edible strain of sweet lupin which has been developed through conventional breeding techniques.

## Biotechnology issues of relevance to developing countries

Biotechnological research as applied to bio-processing in the majority of developing countries, targets development and improvement of traditional fermentation processes. This aspect of the overview considers some areas specifically relevant to developing countries and some key issues that should be considered are listed. Socio-economic and cultural factors Traditional fermentation processes employed in most developing countries are low input, appropriate food processing technologies with minimal investment requirements. They make use of locally produced raw materials and are an integral part of village life. These processes are, however, often uncontrolled, unhygienic and inefficient and generally result in products of variable quality and short shelf lives. Fermented foods, nevertheless, find wide consumer acceptance in developing countries and contribute substantially to food security and nutrition. The question now is: How will applications of biotechnology to fermented foods impact on these socio-economic and cultural factors? Infrastructural and logistical factors Physical infrastructural requirements for the manufacture, distribution and storage (e.g. by refrigeration) of microbial cultures or enzymes on a continuous basis is generally available in urban areas of many developing countries. However, this is not the case in most rural areas of developing countries. The question now is: Should research be oriented to ensure that individuals at all levels can benefit from applications of biotechnology in food fermentation processes, i.e. should logistical arrangements for starter culture development be integrated into biotechnological research targeting improvement of traditional fermentations? What is required for the level of fermentation technologies and process controls to be upgraded in order to increase efficiency, yields and the quality and safety of fermented foods in developing countries?

*Nutrition and food safety*—Fermentation processes enhance the nutritional value of foods through the biosynthesis of vitamins, essential amino acids and proteins, through improving protein and fibre digestibility; enhancing micronutrient bioavailability and degrading anti-nutritional factors. Many bacteria in fermented foods also exhibit functional properties (probiotics). The safety of fermented food products is enhanced through reduction of toxic compounds, such as mycotoxins and cyanogenic glucosides, and production of antimicrobial factors, such as bacteriocins, carbon dioxide, hydrogen peroxide and ethanol, which facilitate inhibition or elimination of food-borne pathogens. The question now is: Are the nutritional characteristics (and safety aspects) of fermented foods adequately documented and appreciated in developing countries? Is there a need for consumer education about the benefits of fermented foods?

*Intellectual property rights (IPRs)*—The processes used in the more advanced areas of agricultural biotechnology tend to be covered by IPRs and these rights tend to be owned by parties in developed countries. This applies also to biotechnology processes used in food processing. On the other hand, many of the traditional fermentation processes applied in developing countries are based on traditional knowledge. In addition to biotechnology processes, microbial strains may also be the object of IPRs. For example, an era of massive private investment in biotechnology was initiated when the United States Supreme Court ruled in 1980 (in the Diamond versus Chakrabarty case) that a live GM bacterium (of the genus *Pseudomonas*, modified to degrade components of crude oil) could be patented.

Many of the microorganisms associated with traditional fermentation processes in developing countries are unique. Issues of ownership will become increasingly important as bacterial strains are characterized and starter cultures are developed in developing countries. The question now is: How should food scientists, researchers, industry and governments in developing countries approach these issues? A considerable volume of research into the development and improvement of fermentation processes is currently taking place worldwide. Are the research results from developing countries adequately documented? Who owns this information? Are cell banks being developed to protect microbial strains characterized in developing countries?

*Commercial opportunities*—Biotechnology also allows the economically viable production of valuable, naturally occurring compounds that cannot be manufactured by other means. For example, commercial-scale production of the natural and highly marketable sweetener known as fructans has long eluded food-processing engineers. Fructans, which are short chains of the sugar molecule fructose, taste like sugar but have no calories. Scientists found a gene that converts 90 percent of the sugar found in beets to fructans. Because 40 percent of the transgenic beet dry weight is fructans, this crop can serve as a manufacturing facility for fructans. Biotechnological innovations have greatly assisted in industrializing production of certain indigenous fermented foods. Indonesian tempe and Oriental soya sauce are well known examples of indigenous fermented foods that have been industrialized and marketed globally. The results of biotechnology research will lead to fermented foods of improved quality, safety and consistency. The question now is: Should biotechnology developments in developing countries target commercialization? Should they target diversification into new value-added products? Should biotechnology development be linked to technological developments in food processing? Can the application of biotechnology to food processing allow farmers in developing countries to add value to their agricultural products (for export or for local consumption) and improve their revenues?

*Appropriateness of biotechnology in developing countries*—As with any commitment of resources, investments in biotechnology should be weighed up against other potential uses of these resources in developing countries. How relevant and worthwhile can such investments on biotechnology be for developing countries?

Biotechnology revolution has spawned new industries focused on manipulating human, animal, plant and micro-Okonko et al. 2365 bial agents to create heretofore unattainable products and services such as fermented food products which have great potential as key protein, fatty acid and good sources of gross energy, therefore, condiments are basic ingredients for food supplementation and their socioeconomic importance cannot be over emphasized in many countries especially in Africa and Asia (India) where protein calorie malnutrition is a major problem. Biological organisms that are being used in most biotechnological processes are microorganisms, which are very small living things which are invisible to the naked eyes but can only be seen with the aid of a microscope. They play this vital role because of the simplicity of their genome, their short generation time, ease of manipulation, their use of synthetic medium for growth among other factors. The present molecular techniques such as cloning, genetic engineering, recombinant DNA technique and polymerase chain reaction (PCR), ribotyping, using restriction fragment length polymorphism (RFLP), and pulsed-field gel electrophoresis (PFGE), Random amplified polymorphic DNA (RAPD) or amplified fragment length polymorphism (AFLP) molecular marker systems involve genetic manipulations using microorganisms such as bacteriophages and bacterial plasmids as vectors and bacterial cells as hosts. This 'therefore' implies that biotechnology occupy a very strategic position in the socio-economic advancement and development of the nation in particular and the world at large.

# CHAPTER-10

# Biotechnology in Some Developing Countries

There are in a number of developing countries – those with advanced research and development in the life sciences – public research institutions and biotechnology companies which have invested in biotechnology R&D (medicine, agriculture and environment), and represent successful stories at national and even regional level.

## Argentina

The family-owned pharmaceutical company Sidus, based in Buenos Aires, made the decision to invest in medical biotechnology by the early 1980s and created a subsidiary, BioSidus, which started producing its first biotechnology-derived product – recombinant erythropoietin (EPO) – by the end of the 1990s. In 1993, BioSidus reached its financial autonomy, i.e. Sidus had been investing money for 13 years in its subsidiary before it became profitable. The investment has been estimated at about $35 million, that can be considered as venture capital responding to a long-term vision for development. Nowadays, BioSidus has a ten-year business plan and much more investment is needed, which probably requires that the company goes public and be quoted on the stock market.

In January 2004, cloned calves Pampa Mansa II and Pampa Mansa III were born and they contained a gene coding for human growth hormone to be produced in their milk. The initial step in obtaining the first generation of clones involve taking fibroblastic cells from calf fetuses and then introducing the fetal cells' nucleus into the cytoplasm of enucleated ovocytes to produce embryos; the fetal cells were transformed to contain the gene for human growth hormone before their nuclei were transferred to enucleated ovocytes. The resulting embryos were transferred into Aberdeen Angus heifers. The procedure led to the birth of Pampa Mansa, which produced milk containing the human growth hormone in 2003.

## Brazil

At the University of Sao Paulo, Fernando de Castro Reinach has been eager to link the public and private sectors of Brazilian science ever since hc completed a Ph.D. at Cornell University Medical School and a postdoc at the Medical Research Council (MRC) Laboratory of Molecular Biology in Cambridge, United Kingdom. In 1990, with two other colleagues, he founded Genomic, one of the first Brazilian companies to perform DNA tests, and one of the first to reach the market with a product. The company became involved in paternity searches and is nowadays one of the largest DNA-testing companies in Brazil. In 1997, F. Reinach, professor at the University of Sao Paulo at the age of 35, was among the first group of seven Brazilian scientists to receive a grant from the Howard Hughes Medical Institute. During the same year, after receiving the approval of the scientific director of the State of Sao Paulo Research Foundation (FAPESP, Fundacao de Amparo a Pesquisa do Estado de Sao Paulo) – the country's third-biggest science and technology funding agency – Reinach put together a proposal to sequence the genome of *Xylella fastidiosa*, a bacterium that destroys $100 million worth of Brazilian citrus every year. In 1999, Reinach decided to launch another private venture, comDominio, an internet hosting service that aimed to bring major sectors of the country online. Reinach worked closely with the former head of Brazil's Central Bank and a major backer of the new venture, which became the country's second-largest hosting service provider. Venture capitalist Paulo Henrique Oliveira Santos, president of Votorantim Ventures, a $300 million company affiliated with Brazil's biggest industrial conglomerate, invested also in comDominio.

## Cuba

This country has spent over 20 years a reported $1 billion building up its bio-industry. Research in genetic engineering for medical biotechnology has started by the early 1980s, the focus being alpha-interferon to treat cancer. At the beginning of the 1990s was created the West Scientific Pole which encompasses 53 institutions and scientific centres, under the leadership of the Center for Genetic Engineering and Biotechnology (CIGB), inaugurated in July 1986. The basic objective of all these institutions is to participate in the country's system of public health care and to contribute to solving public health problems. Cuba has an immunization programme of its whole population with respect to 13 vaccines, an average life-expectancy of 75 years and infant mortality of 6 per 1,000. After the approval of the vaccine against *Haemophilus influenzae*-caused meningitis in October 2003, an hexavalent vaccine against meningitis B, diphtheria, tetanus, whooping cough, poliomyelitis and *H. influenzae* meningitis can now be used in the vaccination programmes in Cuba. The longer-term goal is to produce an heptavalent vaccine to immunize against seven diseases.

## China

Chinese scientific research and development backed by a good economic growth and spurred by the will to catch up with the West, China is heavily investing in science and technology. According to the figures published in the 2003 Overview of Science, Technology and Industry of the Organisation for Economic Cooperation and Development (OECD, Paris), the expenses in research and development (R&D) have reached $60 billion in China in 2001. China has 'therefore' become the world's third science-and-technology power (by this standard) behind the USA and Japan. India which has spent $19 billion for R&D in 2001 is among the top ten countries in the world. The budget of the Chinese Academy of Sciences (CAS) – the most prestigious scientific institution of the country – has more than doubled between 1995 and 2000.

With 743,000 researchers, China has the second-largest research population behind the USA, and this figure does not include the large number of Chinese people studying abroad. In 2000, more than 100,000 were studying in one of the OECD countries, particularly in the USA. In the latter, two-thirds of the foreign students are of Asian origin, while Europeans represent only 17% of this foreign-student population.

## India

In 2004, ten biotechnology-derived drugs were being marketed, four industrial units were manufacturing recombinant anti-hepatitis vaccines locally, and indigenouslyproduced recombinant erythropoietin and granulocyte colony-stimulating factor were also on the market. There are several recombinant drugs and vaccines in advanced production stages.

India has become a world leader in the production of generic drugs. It is nowadays an attractive destination for contract research organizations (CROs) businesses that run trials for pharmaceutical groups. These clinical trials – the approval process for any new pharmaceutical – are time-consuming, expensive and ethically tricky; the task involves recruiting hundreds, often thousands, of sick people to volunteer for the testing of experimental medicines, with unknown side-effects. The aim of carrying out clinical trials in India is to reduce the time and funds needed to turn new molecules into marketable drugs – a process that can take up to 20 years and cost $800 million per drug developed. The overall cost advantage in bringing a drug to market by leveraging India aggressively could be as high as $200 million, because of the availability of large patient populations, access to highly-educated talent and lower cost operations, according to the consultancy firm McKinsey. These developments occur when pharmaceutical companies are beginning to consider transferring parts of their research operations to India. It is attractive because of its many scientists and the fact that it is implementing tougher patent protection. Some executives believe India could become as prominent in pharmaceuticals as it is in information technology.

In clinical trials, India, unlike the USA, offers a huge pool of what the industry calls 'treatment-naive' patients – those who have not been tested with rival drugs. A larger pool of such people may lead to faster patient enrolment in trials and thus to a more rapid drug development. In India, there are about 30 million people with heart diseases, 25 million with type-2 diabetes and 10 million with psychiatric disorders. These widespread supposedly 'rich world' diseases are considered an important target for companies looking to test drugs destined for 'western' consumers.

The world's largest CRO, USA-based Quintiles, began operations in India in 1997 and has recruited 6,400 patients for clinical trials in areas such as psychiatry, infectious diseases and oncology. There were a dozen of CROs that have set up offices in India, up from three in 2001. Mike Ryan, business development manager of New Jersey-based CRO Pharmanet, which has been in India for about a year, stated one of the country's attraction is that patients hold physicians in high esteem. As a result, patient compliance in trials is high – as opposed to the USA, where subjects often drop out to seek second opinions. Cathy White, chief executive of Neeman Medical

International, a USA-based subsidiary of India's Max Health Care group, stated companies could save 20%-30% in drug development costs by outsourcing to India.

Most of these savings come from hiring clinical researchers, nurses and information technology staff at less than a third of 'western' wages. Another factor underpinning the shift to drug testing to India is the recent change in medical research rules. In 2003, Indian health authorities adopted guidelines on 'good clinical practice' in line with global norms. Still, some inside the pharmaceutical industry argue that if patients are illiterate, there are serious ethical issues over their consent in a drug trial. Allan Weinstein, vice-president of clinical research and regulatory affairs with Eli Lilly & Co., has stated "India should not be a place to go just because there are a lot of fresh patients". There must be a likelihood that patients involved in a clinical trial will benefit from the drug.

## Singapore

Although Singapore is not considered a developing country, its development in the area of biotechnology has common features with the way it occurred in the technologically advanced developing countries, e.g. China. Singapore began promoting biotechnology in the early 1980s, attracting Glaxo in 1982. During the following decade, Singapore has been pushing research and development; it set up a Bioprocessing Technology Centre Incubator Unit for start-ups, with fully-equipped laboratories in 1997. Yet Singapore spent and is still spending less as a portion of its gross domestic product on research and development than Japan, South Korea or Taiwan. Also investors are shying away from an industry where products take at least a decade to develop. On the other hand, increased competition is coming from less developed countries like China, India and Malaysia, which are building a bio-industry of their own. Cheap labour in China is drawing jobs away and the government has warned that the unemployment rate, currently at 4.5%, was likely to climb to 5.5% in 2003, its highest rate since 1987, with the economy likely to eke out growth no greater than 1%.

Faced with declining returns in electronics, the industry that helped move Singapore into the ranks of the world's wealthiest nations, the government is throwing its administrative power – and at least $2.3 billion in investments, grants and other incentives – behind an endeavour to become an integrated biotechnology hub. Singapore needs to find a new niche for its economic and social development.

The biotechnology initiative has attracted other big-name manufacturers and research talent. Singapore has had the most success in attracting drug companies with tax holidays and other incentives. Among those with factories there are GlaxoSmithKline, Wyeth, Merck & Co., Inc., Schering-Plough and Pfizer, Inc. On 18 October 2000, the latter announced the future investment of $340.6 million in Singapore in order to build its first plant in Asia; this unit is producing medical ingredients for the manufacture of drugs since 2004 with a staff of 250 persons.

In fact the government has sought to attract global drug makers with a five-year, $1.8- billion programme that included research funding, start-up capital, tax breaks and new facilities. That is why Novartis AG, the world's fifth-biggest drug maker, has decided to join several other rivals in order to make Singapore the site for what was expected to be its biggest drug-manufacturing plant in Asia. Daniel Vasella, Novartis AG's chairman, stated the company had selected Singapore because of its research facilities and political stability. Novartis AG's new research institute, which formally opened on 5 July 2004, will focus on developing drugs against dengue fever and other tropical diseases. The company has agreed that its institute will provide research training to local scientists.

Singaporeans were expected to fill a quarter of staff posts at the research unit. Eli Lilly & Co. and Viacell also opened research institutes the costs of which were partly supported by government grants. Singapore has also opened a biopharmaceutical facility with the goal of developing drugs that could be provided on a contract basis to other companies.

Pharmaceutical production swelled about 50% in 2002, to $5.56 billion, but this industry is less labour-intensive than the electronics industry. To encourage companies to do more than make drugs, the Economic Development Board offers to pay up to 30% of the cost of building research-and-development facilities. In 2000, Singapore declared biotechnology the 'fourth pillar' of its economy and spent approximately $570 million to set up three new biotechnology-research institutes. By the end of August 2003, Singapore was putting the finishing touches on a new $286 million Biopolis medical research-and-development complex, built by JTC Corp., the government industrial park operator. Biopolis comprises a huge underground vivarium to house the rodents needed for the research and is surrounded by a high-technology campus of about 200 ha, complete with condominiums, schools and wireless internet access.

Singapore is close to the Equator and has a lot to offer for the study of tropical disea ses endemic to the region, such as malaria, while its own population is affected by illnesses of affluence like cancer and heart diseases. Its advanced telecommunications 119 infrastructure and abundant computing resources are another attraction. Hence the increasing use of bio-informatics in drug discovery. For stem-cell researchers, Singapore offers one of the world's most liberal legal context. It allows stem cells to be taken from aborted fetuses, and human embryos to be cloned and kept for up to 14 days to produce stem cells. This welcoming conditions attracted Alan Colman in 2002; this scientist helped clone the sheep Dolly in 1996, and moved later on to ES Cell International, a venture between Australian investors and Singapore's Economic Development Board, in order to pursue his medical research.

### USA and other countries

On 9 April 2001, the US president decided that research conducted on embryonic stem cells with federal funding could only deal with existing cell lines derived from embryos produced through *in-vitro* fertilization. On 28 April 2004, over 200 US congress persons, Republicans and Democrats, have requested the US president to make the US regulation on the use of stem cells less stringent. The congress people were fearing that US biologists would leave the USA to work in countries with less drastic legislation. They also highlighted that around 400,000 human embryos were frozen in the USA and could 'therefore' be used as potential sources of new stem-cell lines. By mid-June 2004, a spokesman for the US president said restrictions on embryonic research would not be relaxed, while Massachusetts Senator John Kerry, the Democratic presidential candidate, announced he would overturn the current policy if elected in November 2004.

In Australia, the inaugural Stem Cell Summit in 2003 was held amidst a raging debate on the ethical and legal framework that should underpin stem-cell research in the country. With federal legislation passed, supporting regulatory mechanisms under discussion, and a strengthening commercial sector, the 2003 Summit saw many parts of the system in place for Australia to remain at the forefront of global stem-cell research endeavour.

Polls carried put by Research Australia and Biotechnology Australia give some indication of Australians' opinion. A not-for-profit organization, independent of government, Research Australia's activities are supported by members and donors from leading research organizations, academic institutions, philanthropy, community special interest groups, peak industry bodies, biotechnology and pharmaceutical companies, small businesses and corporate Australia. Research Australia is committed to increasing grass roots awareness of the importance and benefits of health and medical research.

Most people do not have moral acceptance of human cloning (82%), but stem-cell research applied to disease prevention and treatment is supported. The use of adult stem cells has strong support (70%) with a slight majority support for using human stem cells derived from embryos (53%).

Polls showed that the Australian community felt that there is not enough information available to help better understand human stem-cell issues. Approximately two-thirds of the Australians interviewed responded that they were not aware or unsure of the various issues relating to human stem-cell research, and seemed to be less aware / more uncertain about adult stem-cell than embryo stem-cell research. Research Australia's on-line poll results revealed that 60% of respondents did not believe that they had enough information about medical research using stem cells. Only 29% believed that sufficient information was available.

The fact that many Australians are less aware of adult stem-cell research may reflect that many Australians gain their information through the debate presented in the media. The media's coverage largely centred on the issues around embryo stem-cell research as this was the focus of ethical and community concerns. It is critical that research progress be realistically presented and not be overstated – the hype and headlines with no outcomes can lose the public's confidence and possibly impact on the research sector's credibility.

While Australians strongly support government regulation, only one-third of the Australian community felt that the use of human gene technology could be effectively regulated; most people felt that it was either not possible (49%) or did not know (18%). Comparison with Research America's polling on this issue shows a fairly consistent view in Australia and the USA that stem-cell research for fighting disease generally has majority community support. There is strong disapproval of human reproductive cloning and a strong desire for regulatory control.

Regarding the provision of information to the community, polls showed that it should be easily accessible, realistic, accurate, clear and timely. It is also important for regulatory processes to be transparent and well communicated. This needs cooperation between research organizations, academia, private industry, government and agencies such as Biotechnology Australia, consumer groups, the media and broad-based organizations such as Research Australia.

Singapore is actively recruiting people who want to work on the human aspects of biotechnology. China, too, is said to be interested. This shows that not everyone's moral code is shaped by Judeo-Christian ethics and by the Kantian approach to human dignity.

However, China's move into biotechnology has been accompanied by the introduction of biosafety regulations and a modern bioethical framework. This has been stressed by foreign observers such as -, a bioethicist at Bochum University, Germany. In 1998, the Chinese government issued a declaration explicitly banning reproductive cloning. Also research on the human genome is governed by strict rules on sample collection and informed consent.

The Chinese government aims to establish national guidelines governing stem-cell research. Two proposals have been made, both adapted from the British regulations, allowing therapeutic cloning. Enforcement and monitoring will also need improvement. It is true that the collaboration between Chinese scientists (e.g. Peking University Stem Cell Research Centre) and foreign research groups obliges the former to abide by international rules. This is also true for publishing in international journals and attracting overseas investments.

CHAPTER-11

# Modern Biotechnology for Food and Agriculture: Risk and opportunity for Poor

The current debate about the potential utility of modern biotechnology for food and agriculture, and the associated potential risks and opportunities, is focused on the initial applications of such biotechnology in industrial country agriculture. The debate is also intertwined with other concerns such as food safety, animal welfare, industrialized agriculture, and the role of private-sector corporations. At present, there is very little commercial utilization of results from modern biotechnology research in developing countries. As a result, the potential contributions of biotechnology to poverty alleviation and enhanced food security and nutrition in developing countries has received little attention, beyond blanket statements of support or opposition.

A debate based on the best available empirical evidence relevant for poor people in developing countries is urgently needed, to identify the most appropriate ways that molecular biology-based research might contribute to the solution of poor people's problems. These problems and the socioe-conomic context in which they occur are so different from the problems and context of the countries where most of the biotechnology debate currently takes place that the positions and conclusions from the current debate are largely irrelevant for poor farmers and poor consumers in developing countries. Despite this, many of the arguments in the current debate are extrapolated to conclusions about the potential utility for poor countries and poor people. We will attempt to provide input into a more focused debate on the role of modern agricultural biotechnology in developing countries, a debate that should and will be led by people from developing countries themselves.

Small-scale farmers in developing countries are faced with many problems and constraints. Pre and post harvest crop losses due to insects, diseases, weeds and droughts result in low and fluctuating yields, as well as risks and fluctuations in incomes and food availability. Low soil fertility and lack of access to reasonably priced plant nutrients, along with acid, salinated, and waterlogged soils and other abiotic factors, contribute to low yields, production risks, and degradation of natural resources as poor farmers try to eke out a living. They are often forced to clear forest or farm ever more marginal land to cultivate crops. Poor infrastructure and poorly functioning markets for inputs and outputs together with lack of access to credit and technical assistance add to the impediments facing these farmers.

These farmers and other rural and urban poor people suffer from food insecurity and poor nutrition, caused in large measure by poverty and lack of nutritional balance in the diet they can afford. About 1.2 billion people, or one of every five humans, live in a state of absolute poverty, on the equivalent of US$1 a day or less. About 800 million people are food insecure and 160 million pre-school children suffer from energy-protein malnutrition, which results in the death of over 5 million children under the age of five each year.

A much larger number of people suffer from deficiencies of micronutrients such as iron and vitamin A. For example, 2 billion people (one of every three) are anemic, usually as a result of iron deficiency. Food insecurity and malnutrition result in serious public health problems and lost human potential in developing countries. Around 70 percent of poor and food-insecure people reside in rural areas, although poverty and food insecurity appear to be growing in urban areas as urbanization proceeds apace in developing countries. The World Bank forecasts that poverty's center of gravity will remain rural in the early decades of the 21st century.

Most rural poor people depend directly or indirectly on agriculture for their livelihood. Poor people in rural or urban areas spend as much as 50–70 percent of their incomes on food. Low productivity in agriculture is a major cause of poverty, food insecurity, and poor nutrition in low-income developing countries. This is true for urban and rural poor people alike. Low productivity means low incomes for farmers and farm workers, little demand for goods and services produced by poor nonagricultural households in the rural areas, and unemployment and underemployment in urban areas. It also means high unit costs for food, which translate into reduced consumer purchasing power. High food prices are a serious matter for households that spend a large share of their budget on food. In low-income developing countries, agriculture is the driving force for broad-based economic growth and poverty alleviation. A healthy agricultural economy offers farmers incentives for sound management of the natural resource base upon which their livelihood depends.

These relationships are borne out not only by research but also by history in both developing and industrial nations. Productivity increases in European and the U.S.A. agriculture were extremely important to broad-based economic growth during earlier periods of development. More recently, productivity increases in agriculture, led by agricultural research — the Green Revolution — formed the locomotive of rapid broad-based economic growth and poverty reduction in many Asian countries, including China, Indonesia, South Korea, and India. Recent IFPRI research in four African countries found similar strong linkages between agricultural productivity growth and general economic growth.

Productivity gains are essential not only for economic growth and poverty alleviation, but to assure that food supplies remain adequate for a growing world population. According to United Nations projections, world population will increase by 25 percent to 7.5 billion in 2020. On average, 73 million people will be added annually. Over 97 percent of the projected growth will take place in developing countries.

## Public investment critical to food security

Agriculture must figure prominently in poverty alleviation strategies of developing countries. Accelerated public investments are needed to facilitate agricultural and rural growth through:

- yield-increasing crop varieties, including those that are drought and salt tolerant and pest resistant and improved livestock,
- yield-increasing and environmentally friendly production technology,
- reliable, timely and reasonably priced access to appropriate inputs such as tools, fertilizer, and when needed pesticides as well as the credit often needed to purchase them,
- strong extension services and technical assistance to communicate timely information and developments in technology and sustainable resource management to farmers

and to relay farmer concerns to researchers,

- improved rural infrastructure and effective markets,
- particular attention to the needs of women farmers, who grow much of the locally produced food in many developing countries,
- primary education and health care, clean water, safe sanitation and good nutrition for all.

These investments need to be supported by good governance and an enabling policy environment, including trade, macroeconomic, and sectoral policies that do not discriminate against agriculture, and policies that provide appropriate incentives for the sustainable management of natural resources, such as secure property rights for small farmers. Development efforts must engage poor farmers and other low-income people as active participants, not passive recipients; unless the affected people have a sense of ownership, development schemes have little likelihood of success.

Developing countries must reverse present declining levels of public investment in agriculture. On average, they devote 7.5 percent of government spending to agriculture. For their part, donor countries must redress the precipitous decline in aid to agriculture and rural development, which plunged by nearly 50 percent in real terms between 1986 and 1996. Overall development aid has also fallen in recent years. Ironically, our research has found that aid to developing country agriculture not only is effective in promoting sustainable development and poverty alleviation, but it leads to increased export opportunities for industrial countries as well, including, paradoxically, increased agricultural exports. Donors must also rethink their rather inflexible emphasis of the past two decades on less government and a smaller public sector, which has contributed to public disinvestment in agriculture in the developing countries.

## Agricultural research is essential

Public investment in agricultural research is of particular importance for achieving food security in developing countries. The private sector is unlikely to undertake much of the research needed by small farmers because it cannot expect sufficient returns to cover costs. IFPRI research has shown that the annual rates of return to agricultural research and development are, on average, 73 percent. Benefits to society from agricultural research can be extremely large but will not be obtained without public investments. We have also found that even minor increases in aid to agricultural research for developing countries can significantly accelerate food supplies, while relatively small cuts could have serious negative effects.

Despite this evidence, low-income developing countries grossly under-invest in agricultural research; less than 0.5 percent of the value of their agricultural production, compared to 2 percent in higher-income countries. Sub-Saharan Africa, which desperately needs productivity increases in agriculture, has only 42 agricultural researchers per million economically active persons in agriculture, compared with 2,458 in industrial countries.

Efforts to improve longer-term productivity on small-scale farms, with an emphasis on staple food crops, must be accelerated. Research and policies are also needed to help farmers, communities, and governments better cope with risks resulting from such factors as poor market integration, poorly functioning markets, and climatic fluctuations. More research must be directed to the development of appropriate technology for sustainable intensification of agriculture in resource- poor areas, where a high percentage of poor people live, and where environmental risks are severe. The needed research must join all appropriate scientific tools together with better use of the insights of traditional indigenous knowledge.

Research and technology alone will not drive agricultural growth. The full and beneficial effects of agricultural research and technological change will materialize only if government policies are conducive to and supportive of poverty alleviation and sustainable management of natural resources.

## Agricultural biotechnology and food security

Can molecular biology-based research contribute to the solution of the problems outlined earlier? Are the potential social and economic benefits likely to exceed potential risks or costs? If these questions are answered in the affirmative, issues related to the design of the technology and the needed policies and institutions must be tackled. Although conventional applications of biotechnology, such as tissue culture and fermentation amongst others, is under way in several developing countries, little genetically improved (transgenic) seed material has been grown in the poorer developing countries. A great deal is known, however, about the social and economic risks and benefits associated with traditional Mendelian plant breeding as exemplified by the Green Revolution.

The analysis, therefore, begins with the identification of similarities and differences between the Green Revolution and modern biotechnology, and an attempt is made to draw lessons from the Green Revolution and to look at the difference between that technology package and modern biotechnology to try to assess the likely social and economic risks and benefits of modern agricultural biotechnology.

## The green revolution and modern biotechnology

There are three differences of particular importance for an assessment of social and economic risks and benefits. The research leading to the Green Revolution was undertaken by the public sector and the improved seed was usually freely available for seed multiplication and distribution. Although breeders' rights may permit an initial charge for the improved materials, the intellectual property rights (IPR) did not extend beyond the initial release. Having acquired the seed, farmers could reuse it without further payment, although reuse of hybrid seed would drastically reduce the yield advantage. This is in keeping with the principle of 'farmers' rights' included in the 1983 International Undertaking on Plant Genetic Resources.

In contrast, the bulk of modern agricultural biotechnology research is undertaken by private sector firms, which protect IPRs through patents that extend beyond the first release. Farmers, therefore, cannot legally plant or sell for planting the crop produced with the patented seed without the permission of the patent holder. Patent holders, currently seeking ways to enforce their rights, are considering approaches such as legal agreements and technologies that will activate and deactivate specific genes. However, monitoring and enforcing contracts that prohibit large numbers of small farmers from using the crops they produce as seed would be expensive and difficult.

The so-called terminator gene is the first patented technology aimed at biological IPR protection. It is not appropriate for small farmers in developing countries because existing infrastructure and production processes may not be able to keep fertile and infertile seeds apart. Small farmers could face severe consequences if they planted infertile seeds by mistake. Commercialization of the terminator gene now seems unlikely in the short term.

Research is under way on other biological approaches to IPR protection that would not impose such risk on small farmers. These include, for example, genetically engineered seeds that contain desired traits, such as pest resistance or drought tolerance, but in which these are activated only through chemical treatment. Otherwise, the seed would maintain its normal characteristics. Thus, if a farmer planted an improved seed, the offspring would not be sterile; rather they would revert to normal seeds, without the improved traits. The farmer would have the choice of planting the seed and doing no more, or activating the improved traits by applying the chemical. This approach complies with the principle of doing no harm.

It is important to note that even when patents permit a private company to enjoy monopoly or near-monopoly rights over a product it has developed, the firm is unlikely to capture 100 percent of the economic benefits. A recent study of the distribution of the economic benefits generated by the use of herbicide-tolerant soybean seed in the United States in 1997 found that the company, Monsanto, received 22 percent, while seed companies gained 9 percent. Consumers of soybean and soybean products in the United States and other countries reaped a 21 percent share, whereas farmers worldwide obtained 48 percent. The share of U.S. farmers was actually 51 percent of the benefits, but farmers elsewhere experienced net losses of 3 percent.

## Rise of proprietary research processes and technologies

A second, and related, difference between the Green and Gene revolutions involves the patenting of processes as well as products. The main process behind the Green Revolution was conventional plant breeding technology, which lies in the public domain, carried out by public institutions. Today, the processes used in modern agricultural biotechnology are increasingly subjected to IPR protection, along with the products that result. This means that public sector research institutions may not be able to gain access to basic but proprietary knowledge and processes needed in research, including research on the so-called orphan critical staples in the diets of many poor people, but they do not offer promising economic returns to private sector R&D efforts, so efforts to develop disease-resistant cassava or drought-tolerant millet, whether through genetic modification or conventional breeding, must come from the public sector. Some firms have agreed to transfer proprietary technologies, without charging royalties, to developing countries where there are few potential commercial prospects. Monsanto, for example, has entered into agreements with Kenyan and Mexican government agricultural research institutes to develop virus-resistant crops. Arrangements such as these are few and generally involve the philanthropic arms of the private firms.

## Enlightened adaptation vs. direct transfer

A third difference involves the adaptation of industrial country agricultural research to developing country conditions. Although based on earlier research in industrial countries, the Green Revolution was focused on solving specific problems in developing countries. Current application of modern biotechnology is focused on industrial country agriculture.

Industrial country research institutions had begun working on development of higher yielding crop varieties in the late 19th century. For example, in Japan, rice breeding under the auspices of the Ministry of Agriculture and public universities led to large yield gains in the early part of the 20th century, with a second wave of major gains after 1945.

During the early decades of Soviet history, under the leadership of geneticist Nikolai Ivanovich Vavilov, the government carried out extensive crop improvement programs and established one of the world's largest germplasm collections. In the United States, hybrid maize research began in the 1920s. Much of the basic research was done by public institutions, such as land grant universities, state experiment stations, and the U.S. Department of Agriculture (USDA). Applications to particular farming conditions and the mass marketing of the new varieties were, in turn, handled by private seed firms such as Pioneer Hi-Bred and DeKalb.

The research focused not only on developing higher yielding seeds to bolster food supplies for domestic consumption but also on animal feed and production for export. This research could not simply be transferred to poorer developing countries, where the need was for improved varieties of locally-consumed staples. The research that led to the Green Revolution involved further adaptation to the agro-ecological conditions of tropical and semitropical areas. It also focused on rice, wheat, maize, root and tuber crops, and tropical fruits and vegetables. The public sector role was, if anything, even more prominent, with international agricultural research centers (IARCs) and national agricultural research systems (NARS), particularly in Asia and Latin America, playing a prominent role. Financial support came from donors of official development assistance and large private foundations, such as Ford, Rockefeller, and Kellogg.

In contrast, modern agricultural biotechnology is still in an early phase, and the focus is overwhelmingly on production on industrial country farms and for industrial country markets. In 1998, 85 percent of the land planted to genetically improved (GI) crops was in just five developed countries (Australia, Canada, France, Spain, and the United States), with the United States alone accounting for about 75 percent of the area. Argentina, China, Mexico, and South Africa cultivated the remaining 15 percent, and the countries other than China include a substantial number of large-scale, capital-intensive farms that produce primarily for industrial country markets. Among the crops produced in these four developing countries are insect-resistant cotton and maize, herbicide-resistant soybean, and tomatoes with a long shelf life. Globally, herbicide- resistant soybean, insect-resistant maize, and genetically improved cotton (containing insect resistance and/or herbicide tolerance genes) account for 85 percent of all plantings.

Both the area planted to genetically improved crops and the value of the harvests grew dramatically between 1995 and 1999; from less than 1 million hectares to 28 million in 1998 and approximately 40 million in 1999, and from US$75 million in 1995 to US$1.64 billion in 1998. Private industry has dominated research (there are a few exceptions: for example, Rockefeller Foundation support for research on rice, USDA's role in developing the terminator technology, and modest programs at IARCs). Consolidation of the industry has proceeded rapidly since 1996, with more than 25 major acquisitions and alliances worth US$15 billion.

Little private-sector agricultural biotechnology research so far has focused on developing country food crops other than maize. Moreover, little adaptation of the research to developing country crops and conditions has occurred through the 'enlightened' (that is, not for profit, public goods oriented) public and philanthropic channels prominent in the Green Revolution of the developing countries. Some of the exciting international and regional programs are described by Cohen. A program directed at public/private sector linkages is that of the International Service for the Acquisition of Agri-biotech Applications (ISAAA), which transfers and delivers appropriate biotechnology applications to developing countries and builds partnerships amongst institutions.

Relatively little biotechnology research currently focuses on the productivity and nutrition of poor people. The Rockefeller Foundation's agriculture program is one example; in 1998, it provided about US$7.4 million for biotechnology research relevant to developing countries, mainly through IARCs and

NARS in developing countries, with a major emphasis on rice. This sum pales by comparison with Monsanto's 1998 R&D budget of US$1.3 billion, much of which funded agricultural biotechnology research.

As with the Green Revolution, the challenge is to move from the scientific foundation established by industrial country-oriented research efforts to research focused on the needs of poor farmers and consumers in developing countries. Direct transfers of the fruits of agricultural biotechnology research to the developing countries will not work, in most cases. More appropriate research for the developing world might focus on biotechnology and conventional breeding to develop alternative forms of weed resistance, such as leafier rice that denies weeds sunlight rather than incorporating herbicide tolerance into rice. The West Africa Rice Development Association (WARDA), a public IARC in Côte d'Ivoire, has used a combination of conventional plant breeding and tissue culture to develop such rice.

Insect-resistant crops would have great potential value for poor farmers. So far, however, the development of crops containing genes from the Bacillus thuringiensis (Bt ) bacterium, which produces a natural pesticide, has focused largely on the crops and cropping environments of North America. The new crop varieties containing the Bt gene require extremely knowledge-intensive cultivation. They might well be transferable to larger scale operations in some developing countries such as Argentina. The potential usefulness of this application in crops grown by small farmers is open to question. There is considerable debate about risks of the development of resistance in pests, harm to beneficial insects, and crosspollination of wild and weedy plants with the novel gene. The evidence on these issues is still inconclusive and warrants careful monitoring before the application of Bt is tried on a large scale in crops grown by subsistence farmers.

Research on crops and problems of relevance to small farmers in developing countries will require the allocation of additional public resources to agricultural research, including biotechnology research, that promises large social benefits. There is no reason to believe that this research will offer lower rates of return than other agricultural research and development. Private-sector agricultural research currently accounts for a small share of agricultural research in most developing countries. The public sector can expand private-sector research for poor people by converting some of the social benefits to private gains, for example, by offering to buy exclusive rights to newly developed technology and make it available either for free or for a nominal charge to small farmers. The private research agency would bear the risks, as it does when developing technology for the market. IARCs have an important role to play as intermediaries in facilitating such arrangements.

Without more enlightened adaptation, continued expansion of genetically improved crop production in the industrial countries may well have a negative impact on small farmers in developing countries. Some developing country consumers would benefit, but those consumers who also farm could experience net losses. In addition, the development of industrial substitutes for developing country export crops, such as cocoa (which in many developing countries is produced by small farmers) could have a devastating impact on developing country farmers' livelihoods. In sum, the biggest risk of modern biotechnology for developing countries is that technological development will bypass poor farmers and poor consumers because of a lack of enlightened adaptation.

It is not that biotechnology is irrelevant, but that research needs to focus on the problems of small farmers and poor consumers in developing countries. Private sector research is unlikely to take on such a focus, given the lack of future profits. Without a stronger public sector role, a form of 'scientific apartheid" may well develop, in which cutting edge science becomes oriented exclusively toward industrial countries and large-scale farming.

## Lessons from the Green Revolution

The outcomes of the Green Revolution offer some guideposts for assessing the likely risks and benefits of agricultural biotechnology for developing countries. Risks and benefits may be inherent in a given technology, or they may transcend the technology. The policy environment into which a technology is introduced is critical. For example, IFPRI research has found that in Tamil Nadu State in India, the adoption of high-yielding grain varieties meant not only increased yields and cheaper, more abundant food for consumers, but income gains for small and largerscale farmers alike, as well as for nonfarm poor rural households. Increased rural incomes contributed to nutrition gains for these households.

Because the Tamil Nadu state government has pursued active poverty alleviation strategies, including extensive social safety net programs and investment in agriculture, rural development, and a fair measure of equity in access to resources such as land and increased inequality followed the adoption of Green Revolution technology, it was not because of factors inherent to the technology, but rather a result of policies that did not promote equitable access to resources. And even in these areas, rural landless labourers usually found new job opportunities as a consequence of increased agricultural productivity, particularly where appropriate physical infrastructure and markets developed. Successful adoption of Green Revolution technology, however, depended on access to water, fertilizer and pesticides. Thus, inequality between well-endowed and resource-poor areas increased because of the properties of the technology itself. Likewise, excessive or improper use of chemical inputs led to adverse environmental impacts in some instances. This problem was offset, to some extent, by characteristics that were also inherent in the technology; by allowing yield gains without expanding cultivated area, the technology kept cultivators from clearing forests and moving onto wild and marginal lands.

Overall, the Green Revolution was extremely successful in enhancing productivity in rice, wheat and maize; in increasing incomes and reducing poverty; and in preserving forests and marginal lands by improving yields within existing cultivated areas. By reducing unit costs and prices for food, it greatly benefited poor consumers and by boosting farmers' incomes it contributed to gains in nutrition. Would agricultural biotechnology produce similar results in developing countries? The answer depends on whether the research is relevant to poor people and on its ownership, that is, the nature of the intellectual property rights arrangements.

## Weighing Risks and Benefits of Biotechnology

Modern biotechnology is not a silver bullet for achieving food security, but, used in conjunction with traditional or conventional agricultural research methods, it may be a powerful tool in the fight against poverty that should be made available to poor farmers and consumers. It has the potential to help enhance agricultural productivity in developing countries in a way that further reduces poverty, improves food security and nutrition, and promotes sustainable use of natural resources. Solutions to the problems facing small farmers in developing countries will benefit both farmers and consumers.

The benefits of new genetically improved food to consumers are likely to vary according to how they earn their income and how much of their income they spend on food. Consumers outnumber farmers by a factor of more than 20 in the European Union, and Europeans spend only a tiny fraction of their incomes on food. Similarly, in the United States, farms account for less than 2 percent of all households, and the average consumer spends less than 12 percent of income on food. In the industrial countries, consumers can afford to pay more for food, increase subsidies to agriculture, and give up opportunities for better-tasting and better-looking food. In developing countries, poor consumers depend heavily on agriculture for their livelihoods and spend the bulk of their income on food. Strong opposition to GI foods in the European Union has resulted in restrictions on modern agricultural biotechnology in some countries.

The opposition is driven in part by perceived lack of consumer benefits, uncertainty about possible negative health and environmental effects, widespread perception that a few large corporations will be the primary beneficiaries, and ethical concerns.

## Potential benefits

There are many potential benefits for poor people in developing countries. Biotechnology may help achieve the productivity gains needed to feed a growing global population, introduce resistance to pests and diseases without costly purchased inputs, heighten crops' tolerance to adverse weather and soil conditions, improve the nutritional value of some foods and enhance the durability of products during harvesting or shipping. New crop varieties and biocontrol agents may reduce reliance on pesticides, thereby reducing farmers' crop protection costs and benefiting both the environment and public health. Biotechnology research could aid the development of drought-tolerant maize and insect-resistant cassava, to the benefit of small farmers and poor consumers. Research on genetic modification to achieve appropriate weed control can increase farm incomes and reduce the time women farmers spend weeding, allowing more time for the child care that is essential for good nutrition. Biotechnology may offer cost-effective solutions to micronutrient malnutrition, such as vitamin A- and iron-rich crops.

Research focused on how to reduce the need for inputs and increase the efficiency of input use could lead to the development of crops that use water more efficiently and extract phosphate from the soil more effectively. The development of cereal plants capable of capturing nitrogen from the air could contribute greatly to plant nutrition, helping poor farmers who often cannot afford fertilizers. By raising productivity in food production, agricultural biotechnology could help further reduce the need to cultivate new lands and help conserve biodiversity and protect fragile ecosystems. Productivity gains could have the same poverty-reducing impact as those of the Green Revolution if the appropriate policies are in place. Policies must expand and guide research and technology development to solve problems of importance to poor people. Research should focus on crops relevant to small farmers and poor consumers in developing countries, such as ba-nana, cassava, yam, sweet potato, rice, maize, wheat, and millet, along with livestock.

## Health and environmental risks

Genetically improved (GI) foods are not intrinsically good or bad for human health. Their health effect depends on their specific content. GI foods with a higher iron content are likely to benefit iron-deficient consumers. But the transfer of genes from one species to another may also transfer characteristics that cause allergic reactions. Thus, GI foods need to be tested for allergy transfers before they are commercialized. Such testing avoided the possible commercialization of soybeans with a Brazil nut gene. GI foods with possible allergy risks should be fully labeled.

Labeling may also be needed to identify content for cultural and religious reasons or simply because consumers want to know what their food contains and how it was produced. While the public sector must design and enforce safety standards as well as any labeling required to protect the public from health risks, other labeling might best be left to the private sector in accordance with consumer demands for knowledge

Failure to remove antibiotic-resistant marker genes used in research before a GI food is commercialized presents a potential although unproven health risk. Recent legislation in the

European Union requires that these genes be removed before a GI food is deemed safe. Risks and opportunities associated with GI foods should be integrated into the general food safety regulations of a country. International agencies and donors may need to assist some developing countries build the capacity to develop appropriate regulatory arrangements.

These regulatory systems are needed to govern food safety and assess any environmental risks, monitor compliance, and enforce such regulations. The regulatory arrangements should be country-specific and reflect relevant risk factors. Progress on achieving a global agreement on biosafety standards is urgently needed. The development of a public global regulatory capacity has lagged far behind the pace of economic globalization. The ecological risks policymakers and regulators need to assess include the potential for spread of traits such as herbicide resistance from genetically improved plants to unmodified plants (including weeds), the buildup of resistance in insect populations, and the potential threat to biodiversity posed by widespread monoculture of genetically improved crops. Seeds that allow farmers the option of 'turning off' genetic characteristics, mentioned earlier, offer great promise for assuring that new traits do not spread through cross-pollination.

Both food safety and biosafety regulations should reflect international agreements and a given society's acceptable risk levels, including the risks associated with not using biotechnology to achieve desired goals. Poor people should be included directly in the debate and decision-making about technological change, the risks of that change, and the consequences of no change or alternative kinds of change.

## Socioeconomic risks

Unless developing countries have policies in place to ensure that small farmers have access to delivery systems, extension services, productive resources, markets, and infrastructure, there is considerable risk that the introduction of agricultural biotechnology could lead to increased inequality of income and wealth. In such a case, larger farmers are likely to capture most of the benefits through early adoption of the technology, expanded production, and reduced unit costs.

Growing concentration among companies engaged in agricultural biotechnology research may lead to reduced competition, monopoly or oligopoly profits, exploitation of small farmers and consumers, and extraction of special favors from governments. Effective antitrust legislation and enforcement institutions are needed, particularly in small developing countries where one or only a few seed companies operate. Global standards regarding industrial concentration must also be developed; international public policies in this area have not kept pace with economic globalization. Effective legislation is also required to enforce IPRs, including those of farmers to germplasm, along the lines agreed to within the WTO and the Convention on Biological Diversity.

## Ethical questions

A major ethical concern is that genetic engineering and 'life patents' accelerate the reduction of plants, animals and microorgan-isms to mere commercial commodities, bereft of any sacred character. This is far from a trivial consideration. However, all agricultural activities constitute human intervention into natural systems and processes, and all efforts to improve crops and livestock involve a degree of genetic manipulation. Continued human survival depends on precisely such interventions.

Expanded enlightened adaptive research on agricultural biotechnology can contribute to food security in developing countries, provided that it focuses on the needs of poor farmers and consumers in those countries, identified in consultation with poor people themselves. It is also critical that biotechnology be viewed as one part of a comprehensive sustainable poverty alleviation strategy, not a technological quick-fix for world hunger. Biotechnology needs to go hand in hand with investment in broad-based agricultural growth. There is considerable potential for biotechnology to contribute to improved yields and reduced risks for poor farmers, as well as more plentiful, affordable, and nutritious food for poor consumers. It is not, as some critics have charged, "a solution looking for a problem." The problems are genuine and momentous. Public sector research, is essential for ensuring that molecular biology-based science serves the needs of poor people. It is also urgent that internationally accepted biosafety standards and local regulatory capacity be strengthened within developing countries.

Evaluation of genetically improved crops needs to increase in developing countries; at present, about 90 percent of the field testing occurs in industrial countries. Without field testing, it is virtually impossible to assess potential, environmental and health risks. Hence, destruction of test plots by anti-GI activists should cease. Open debate about the issues involved is essential, but physical attacks on research and testing efforts contribute little to the free exchange of ideas or the formulation of policies that will advance food security.

If the appropriate steps, including those outlined above, are not taken, modern biotechnology could bypass poor people. Opportunities for reducing poverty, food insecurity, child malnutrition and natural resource degradation will be missed, and the productivity gap between developing and industrial country agriculture will widen. Such an outcome would be unethical indeed.

CHAPTER-12

# Genetic Resource Policies

The introduction of biotechnology into the agri-food world in the 1990s complicated an already difficult regulatory and trade system. At one level, biotechnology and genetically modified (GM) foods increase the potential for trade and the need for a fully functioning international trading system. At another level, the products of this new technology have precipitated a large and difficult debate about the structure and effectiveness of national food safety regulations and the appropriate role for international institutions. A number of national and international efforts are underway to manage these pressures, but prospects for early resolution are not great.

Biotechnology is inextricably linked to international trade. The technology has been globally developed and is being applied to research programs in more than 30 countries around the world. Biotechnology has had the greatest effect on the most heavily traded agri-food commodities in the global trading system. Although the first biotechnology-based agri-food product entered the market only in 1994, by 2001 more than 50 modifications involving 13 crops had been approved and produced on more than 52 million hectares in at least 14 countries. Commercial production of GM foods has been concentrated in canola, corn, cotton, and soybeans, which are extensively traded internationally. Perhaps most important, GM production has been concentrated in countries that are the traditional and dominant exporters of those crops (particularly Argentina, Canada, China, and the United States). Up to 88 percent of trade in some of the products with GM varieties comes from the key GM-adopting countries.

For the most part, GM products have been marketed as commodities and mixed with batches of GM and non-GM products as they flow into the international marketplace and then to many countries around the globe. Once in these markets, the commodities are extensively processed, and their components (edible oils, corn meals, soybean proteins, and so on) are fundamental ingredients in more than 70 percent of the processed foods available in most developed-country markets. GM products appear to simply raise new concerns about access to international markets. Those few countries producing and exporting the products seek to be able to continue their business unimpeded.

Yet GM varieties tend to exacerbate the debate about market access because almost all the biotechnology traits in commercial production—herbicide tolerance, insect resistance, and viral resistance—lower production costs or increase yields. Those countries adopting these technologies, which also tend to be traditional exporters, thereby increase their exportable surpluses, depressing world prices and making nonadopting importing producers less competitive. As a result, disadvantaged farmers may join with consumers in importing countries concerned about the safety of these products in calling for increased controls on these products.

### The domestic regulatory response

A number of factors have made this issue hard to handle. Uncertainty about the food and environmental safety of new GM foods has led to different responses in different markets. Those markets, lacking domestic regulators that command the confidence of consumers, have tended to act in a 'precautionary' way, either reviewing the products more slowly or imposing temporary bans on the introduction of the new products. This is a sharp break from the international food safety system that evolved over the past 100 years, where importers tended to accept the food and environmental safety judgments of regulators from those countries developing and exporting the products. One result of this 'renationalization' of agri-food safety regulation is that national systems have tended to diverge. Canada, Japan, Mexico, and the United States, among others, generally make similar rulings and have approved most of the new GM products for production and consumption. Regulators in Australia, the European Union (EU), and New Zealand, in contrast, have postponed approvals in recent years, reflecting the concerns of their citizens. Another 20 or so countries have developed domestic regulatory systems consistent with one or other of these approaches.

The diverging domestic systems are most evident when one looks at the labeling systems being proposed or developed in various countries. So far more than 26 countries have either adopted provisions or announced plans for rules to help the market develop and deliver labeled products. At one extreme, Argentina, Canada, Hong Kong, and the United States have adopted a voluntary labeling strategy that will likely allow labels for either GM or GM-free products, with only 1–5 percent tolerances. At the other extreme, 22 countries and the EU have adopted or announced plans to implement mandatory labeling systems. As of June 2002, only a handful of these countries had revealed the full structure of the labeling rules they intend to pursue, and only Australia, China, Japan, New Zealand, South Korea, and the United Kingdom have formally implemented labeling systems. A number of other countries have proposed mandatory labeling (for example, Brazil, Czech Republic, Hungary, Indonesia, Poland, Russia, South Africa, and Thailand), but there is little available evidence that these countries have developed domestic systems to manage such regulations or, for that matter, any firm indication of when their systems might be operational.

The key concern about the diverging domestic regulatory systems is that production and trade are shifting. Key GM adopters, especially Canada and the United States, are abandoning or losing key markets and diverting their exports to new markets. U.S. exports of corn to the EU have fallen by 70 percent in recent years, U.S. exports of soybeans to the EU have dropped by 48 percent, and Canadian exports of canola to the EU have dropped 96 percent.

Meanwhile, the EU has developed new GM-free sources of soybeans from Brazil and canola from Australia, both markets that have not yet approved GM varieties for those crops. So far these changing trade flows have not significantly affected producer returns—trade has simply been reallocated between adopting and nonadopting countries—but over time such policies have the potential to seriously distort trade flows and offset many of the benefits of recently negotiated international trade agreements for these products.

Most of the rest of the countries in the world do not have any domestic regulatory capacity and are seeking guidance and help from international institutions.

## The international regulatory

Nine international bodies are currently vying to coordinate and regulate different aspects of food safety. These institutions fall into three types. Five are largely science-based organizations: the International Plant Protection Convention (IPPC), International

Epizootics Organization (OIE), Codex Alimentarius (Codex), the Food and Agriculture Organization of the United Nations (FAO), and the World Health Organization (WHO).

One, the World Trade Organization (WTO), is a trade-based organization. The three others have broader objectives such as environmental protection and other social or political goals: the Organisation of Economic Co-operation and Development (OECD), Regional Initiatives, and the Cartagena BioSafety Protocol (BSP). These organizations seek to develop standards for health, safety, and labeling for GM foods, establish testing procedures to ensure the standards are met, provide rules for allowable policies, and create systems to manage disputes.

Despite the substantial effort being undertaken, there is no common view on the goal of international regulation. While most agree that safety is the bottom line, few can agree on what that means, whose opinions should hold the most weight or how to handle non safety issues such as social, economic or ethical concerns. The FAO and WHO have a long history of multilateral efforts to promote food security and public health and have worked to develop a consensus about the implications of biotechnology for their areas of interest. Meanwhile, the IPPC and OIE are multilateral treaties that seek to protect plants and animals from the spread of pathogens through international trade, thereby providing much of the scientific consensus that underlies domestic food safety systems. Both institutions have their own nonbinding dispute avoidance and settlement systems, but their most important role in international trade is through the WTO Sanitary and Phytosanitary Agreement (SPS), which uses the IPPC and OIE standards as the basis for evaluating SPS disputes. National measures based on international standards from either of these institutions will generally not be open to challenge under the WTO dispute resolution process.

**Conserving genetic resources for agriculture**

As improved crop varieties, developed by scientific breeding, spread throughout the world in the latter half of the 20th century, the risk of excessive reliance by farmers and breeders on a narrowing genetic base was dramatized by the infestation and vulnerability of U.S. hybrid corn with cytoplasm male sterility to southern corn leaf blight.

Events like this spurred worldwide efforts to greatly expand the amount of agricultural biodiversity conserved in gene-banks. More recently, microarray and other modern biotechnologies that provide new and less costly ways of screening crop samples for useful traits have increased the value of conserved genetic resources and focused worldwide attention on access to and use rights of traditional crop.

The 11 gene-banks maintained by the research centers of the Consultative Group on International Agricultural Research (CGIAR) conserve more than 6,66,000 accessions (plant or seed samples) of crops grown mainly by poor people, staple food crops grown worldwide and tree species used in agroforestry systems. This collection constitutes a sizable share—perhaps 30 percent or more—of the unique entries in gene-bank collections worldwide. Conservation of this valuable germplasm should have a very long-term, if not perpetual, perspective.

But funding for this long-term conservation service is currently provided on a precarious, year-by-year basis. This mismatch between the generally short-term nature of the financial support and the long-term nature and intent of the effort could threaten the security and future availability of this genetic material. A plan to judiciously match the duration of the funding commitments to the duration of the conservation commitments was unveiled at the World Food Summit in Rome in June 2002 and further elaborated at the World Summit on Sustainable Development in Johannesburg in August 2002. It involves an effort to tap private and public sources of support to establish a Global Conservation Trust (GCT) fund designed to sustain the long-term conservation and use of agricultural germplasm held in ex situ gene-banks.

But just how costly is it to conserve genetic resources in gene-banks and maintain their viability and sample sizes in perpetuity? In this brief we estimate the costs of conserving specific crop species in ex situ gene-banks in perpetuity, including the costs of maintaining healthy and viable seeds and other plant breeding material (collectively called 'germplasm') stored in the field or in vitro. We also show how these estimates change in response to variations among crops, conservation protocols, and institutional arrangements.

The present value of these in-perpetuity costs indicates the necessary size of an endowment or trust fund that would furnish an income stream sufficient to underwrite long-term conservation efforts, thus keeping this valuable resource available for use in maintaining biodiversity and supporting plant breeding for the foreseeable future.

CHAPTER-13

# Life Sciences and Biotechnology

Life sciences and biotechnology are widely recognised to be, after information technology, the next wave of the knowledge-based economy, creating new opportunities for our societies and economies. They also raise important policy and societal issues and have given rise to a broad public debate. These issues must be addressed with great care and sensitivity.

In Europe, however, the relevant responsibilities fall across a broad range of policies and actors. In the absence of a shared vision of what is at stake and without common objectives and effective coordination, Europe has, therefore, only slowly and with difficulty addressed the challenges and opportunities of these new technologies.

Our democratic societies should offer the necessary safeguards and channels of dialogue to ensure that the development and application of life sciences and biotechnology take place respecting the fundamental values recognised by the EU in the Charter of Fundamental Rights.

Europe is faced with a major policy choice; either accept a passive and re-active role, and bear the implications of the development of these technologies elsewhere, or develop pro-active policies to exploit them in a responsible manner, consistent with European values and standards. The longer Europe hesitates, the less realistic this second option will be

## Technology revolution and policy response

A revolution is taking place in the knowledge base of life sciences and biotechnology, opening up new applications in health care, agriculture and food production, environmental protection, as well as new scientific discoveries. This is happening globally. The common knowledge base relating to living organisms and ecosystems is producing new scientific disciplines such as genomics and bioinformatics and novel applications, such as gene testing and regeneration of human organs or tissues. These in turn offer the prospect of applications with profound impacts throughout our societies and economies, far beyond uses such as genetically modified plant crops.

The expansion of the knowledge base is accompanied by an unprecedented speed in transformation of frontier scientific inventions into practical use and products and thus also represents a potential for new wealth creation; old industries are being regenerated and new enterprises are emerging, offering the kind of skill-based jobs that sustain knowledge-based economies. As probably the most promising of the frontier technologies, life sciences and biotechnology can provide a major contribution to achieve the European Community's Lisbon Summit's objective of becoming a leading knowledge-based economy.

## The potential of life sciences and biotechnology

Life sciences and biotechnology are widely regarded as one of the most promising frontier technologies for the coming decades. Life sciences and biotechnology are enabling technologies - like information technology, they may be applied for a wide range of purposes for private and public benefits. On the basis of scientific breakthroughs in recent years, the explosion in the knowledge on living systems is set to deliver a continuous stream of new applications.

There is a huge need in global health care for novel and innovative approaches to meet the needs of ageing populations and poor countries. There are still no known cures for half of the world's diseases, and even existing cures such as antibiotics are becoming less effective due to resistance to treatments. Biotechnology already enables cheaper, safer and more ethical production of a growing number of traditional as well as new drugs and medical services (e.g. human growth hormone without risk of Creutzfeldt-Jacobs is ease, treatment for haemophiliacs with unlimited sources of coagulation factors free from AIDS and hepatitis C virus, human insulin, and vaccines against hepatitis B and rabies). Biotechnology is behind the paradigm shift in disease management towards both personalised and preventive medicine based on genetic predisposition, targeted screening, diagnosis, and innovative drug treatments.

Pharmacogenomics, which applies information about the human genome to drug design, discovery and development, will further support this radical change. Stem cell research and xenotransplantation offer the prospect of replacement tissues and organs to treat degenerative diseases and injury resulting from stroke, Altzheimer's and Parkinson's diseases, burns and spinal-cord injuries. In the agro-food area, biotechnology has the potential to deliver improved food quality and environmental benefits through agronomically improved crops.

Since 1998, the area cultivated with genetically modified (GM) crops worldwide has nearly doubled to reach some 50 million hectares in 2001 (in comparison with about 12.000 hectares in Europe). Food and feed quality may be linked to disease prevention and reduced health risks. Foods with enhanced qualities (functional foods) are likely to become increasingly important as part of life-style and nutritional benefits. Plant genome analysis, supported by a fair research project, has already led to the genetic improvement of a traditional European cereal crop (called Spelt) with an increased protein yield (18%) which may be used as an alternative source of protein for animal feed.

Considerable reductions in pesticide use have been recorded in crops with modified resistance. The enhancement of natural resistance to disease or stress in plants and animals can lead to reduced use of chemical pesticides, fertilisers and drugs, and increased use of conservation tillage - and hence more sustainable agricultural practices, reducing soil erosion and benefiting the environment. Life sciences and biotechnology are likely to be one of the important tools in fighting hunger and malnutrition and feeding an increasing human population on the currently cultivated land area, with reduced environmental impact.

Biotechnology also has the potential to improve non-food uses of crops as sources of industrial feedstocks or new materials such as biodegradable plastics. Plant-based materials can provide both molecular building blocks and more complex molecules for the manufacturing, and energy and pharmaceutical industries. Modifications under development include alterations to carbohydrates, oils, fats and proteins, fibre and new polymer production. Under the appropriate economic and fiscal conditions, biomass could contribute to alternative energy with both liquid and solid biofuels such as biodiesel and bioethanol as well as processes such as bio-desulphurisation. Plant genomics also contributes to conventional improvements through the use of marker-assisted breeding.

New ways to protect and improve the environment are offered by biotechnology including bioremediation of polluted air, soil, water and waste as well as development of cleaner industrial products and processes, e.g. based on use of enzymes (biocatalysis).

## Harvesting the potential

The potential of life sciences and biotechnology is being exploited at an accelerating rate and is likely to engender a new economy with creation of wealth and skilled jobs. Less certain is the time profile and orientations of this development and whether Europe will fully participate.

Some estimates suggest that by the year 2005 the European biotechnology market could be worth over € 100 billion. By the end of the decade, global markets, including sectors where life sciences and biotechnology constitute a major portion of new technology applied, could amount to over € 2000 billion. Europeans are also likely to become major beneficiaries of solutions offered by life sciences and biotechnology - in the form of products and services for consumers, for public benefits and throughout the production system. But to manage this development, to give us options, to project our values and policy choices internationally, and to reap the benefits of a new emerging economy, Europe should also command the knowledge base and its transformation into new products, processes and services.

## The knowledge-base

The life sciences revolution was born in, and is fed and nurtured by, research. Public research laboratories and institutions of higher education are at the core of the science base interacting also with enterprise-based research and that of other private bodies. The success of any knowledge-based economy rests upon the generation, diffusion and application of new knowledge. Investments in research and development, education and training and new managerial approaches are, therefore, of key importance in meeting the challenges posed by life sciences and biotechnology.

One of Europe's main strengths is its science base; centres of scientific excellence in specific technologies exist and are at the core of regional clusters of biotechnology development. However, total European investment in R&D is lagging behind the US.

Moreover, Europe suffers from fragmentation of public research support, and from the low level of interregional co-operation in R&D, among companies and institutions from different regions of several states. It was aimed to restore European leadership in life sciences and biotechnology research. The 6th Community Framework Programme for Research, Technological development and Demonstration activities propose this area as the first priority and will provide a solid platform for constructing, in collaboration with the Member States, a European Research Area. This should reinforce R&D capacity and help overcome existing fragmentation of research policies and efforts. When Europeans work together, maximising collaboration and minimising duplication, we can better meet major challenges such as the handling of the ever-increasing volumes of data and information and ensuring full participation in global scientific initiatives.

Moreover, European research efforts should focus on the new prospects that are opening up through multi-disciplinary research. New discoveries are made most often in when biological research is carried out in conjunction with other sciences and disciplines such as information technology, chemistry and process engineering. For example, human genome analysis into so-called 'gluten allergy' may ultimately lead to the development of allergen-reduced cereals. A first fully integrated community project has recently been launched to ensure leadership at the genomes medicine interface where biotechnology is yielding innovative approaches to treatments of human and animal diseases.

Europe's research agenda for life sciences should be based on the needs of its citizens and attuned to our particular requirements. This calls for an approach which actively identifies the needs and opportunities presented by European societies and seeks to address them through innovative research. We need to further strengthen the links between research and other community policies, including the scientific basis for health and safety regulations. In the same logic, it is also of utmost importance to involve scientists and researchers as closely as possible in societal consensus building. New research partnerships should also be encouraged amongst developed and developing nations to take full advantage of promising technologies and biodiversity potential, the basis for future progress.

## Societal scrutiny and dialogue

Life sciences and biotechnology have given rise to significant public attention and debate. This public debate was welcomed as a sign of civic responsibility and involvement. Life sciences and biotechnology should continue to be accompanied and guided by societal dialogue.

Dialogue in our democratic societies should be inclusive, comprehensive, well informed and structured. Constructive dialogue requires mutual respect between participants, innovative approaches, and time. It should be structured in agreement with stakeholders to allow progress, for example in the provision of better information and mutual understanding. Experience also shows how important it is that dialogue takes place at the local and national levels, as well as internationally, and the member states and local actors are invited to take relevant initiatives.

Dialogue should be open for all stakeholders. Public authorities should help to ensure participation by stakeholders with limited resources. Economic operators, industry and users, who have economic interests at stake, as well as the scientific community, bear a particular responsibility for active participation. The Commission invites these parties to respond to public concerns, for example through transparency of their visions, policies and ethical standards.

Relevant public information is essential for meaningful dialogue. Providing it requires focused and pro-active efforts. It is especially important that the information needs formulated by the broad public are taken seriously and responded to. We shall also strive for a balanced and rational approach, distinguishing between real issues, on which we must act, and false claims.

**Ethical values and societal goals to be in harmony**

Without broad public acceptance and support, the development and use of life sciences and biotechnology in Europe will be contentious, benefits will be delayed and competitiveness will be likely to suffer.

The debate and the public consultation carried out indicate that the European public is quite prepared and capable to enter into complex weighting of benefits against disadvantages, guided by fundamental values. Although sometimes polarised, the public debate demonstrates many points of converging views. Public opinion depends crucially on the perceived benefits of life sciences and biotechnology. Eurobarometer surveys reveal that public expectations of biotechnology, apart from medical advances, are moderate. And there is also considerable public uncertainty about some applications, and aversion towards their distributional impacts and the risks involved.

There is broad support for many guiding values and goals. Some of these, such as the freedom of research, intrinsic value of new knowledge and the moral obligations to help alleviate illness or hunger, tend to favour the development and application of these new technologies. Others help to clarify the criteria and conditions for the development and applications of life sciences and biotechnology, in particular the need to take into account the ethical and societal implications, and the importance of transparency and accountability in decision-making, minimising risk, and freedom of choice.

It is, therefore, of key importance to support information and dialogue to help the public and stakeholders better understand and appreciate these complex issues and to develop methods and criteria for assessing benefits against disadvantages or risks, including the distribution of impacts among different parts of society. Our democratic societies should offer the necessary safeguards to ensure that the development and application of life sciences and biotechnology take place respecting the fundamental values recognised by the EU in the Charter of Fundamental Rights, in particular by confirming the respect for human life and dignity. The Community has also banned funding of research into human reproductive cloning.

Support should be given to the Frano-German initiative, addressed to the UN, for a world-wide convention on the prohibition of human reproductive cloning. Other issues such as stem cell research clearly require attention and further debate. Europe has also taken clear positions on the importance of freedom of choice for consumers as well as for economic operators with respect to GM foods, and we have established broad societal agreement on the need to safeguard European agricultural practices. However, scientific and technological progress will continue to give rise to new ethical or societal implications. The Commission considers that these issues should be addressed pro-actively and with a broad perspective, taking into account the moral obligations towards present and future generations and the rest of the world. We should not content ourselves with acting defensively only when our core values are being transgressed.

These issues cannot be adequately addressed within the narrow context of regulatory product approvals but require more flexible and forward-looking approaches. Europe needs an active and on-going public dialogue, accompanied by focused fact-finding on both benefits and disadvantages to allow the public to contribute to the complex process of setting priorities. In the context of its Science & Society initiative, a series of actions intended to strengthen the ethical dimension in sciences and new technologies has already been proposed.

CHAPTER-14

# Global Impact of Modern Biotechnology

This chapter debates over the evolution of intellectual property rights laws in India as they relate to agricultural biotechnology. This has been an area subject to significant developments over recent years, and one in which many new and emerging norms remain untested.

Scientific advancements in agricultural biotechnology, aiming at enhancement of food production have been expected to contribute to making food more nutritious and improve the food security position of India. They are also expected to increase the productivity of land and thus reduce the pressure to extend the acreage under cultivation. However fears have been expressed that the benefits of biotechnology could be outweighed by the risks involved. One of the main fears is that affording intellectual property protection to agricultural biotechnology would drastically affect the economic situation of farmers to the extent of endangering food security.

However, the divergence of various interests has hindered the formulation of a clear policy on protection of biotechnological innovation. Realising the 'successes in agriculture...., biotechnology...' and 'major national achievements' in terms of 'very significant increase in food production', science and technology leading to 'new norms of intellectual property rights', and in *The Science and Technology Policy 2003*, the Government of India has declared as its policy objectives:

- to ensure food, agricultural, nutritional, environmental, water, health and energy security of the people on a sustainable basis,
- to mount a direct and sustained effort on the alleviation of poverty, enhancing livelihood security, removal of hunger and malnutrition,
- to establish an Intellectual property Rights regime which maximises the incentives for the generation and protection of intellectual property by all types of inventors. The regime would also provide a strong, supportive and comprehensive policy environment for speedy and effective domestic commercialisation of such inventions so as to be maximal in the public interest.

The remainder of this introduction provides some historical context for recent developments.

## I Food security

The aim of agricultural policies has been to provide food for all. The concept of food security has undergone a shift from the emphasis on availability of food to access to food. If food security is the aim, intellectual property rights which primarily determine economic accessibility of food crops assume paramount importance. Indian policymakers are now 'therefore' being forced to pay greater consideration to the economic management of food security rather than merely the technical aspects. To understand this shift in perspective we will have to go back to the green revolution.

## Green revolution

The government's policy adopted during the green revolution in India wherein public funded research institutions were involved in extensive research, gave an impetus to farming communities to adopt the new technology for agricultural development. The green revolution markedly reduced the costs of farming. The development of High Yielding Varieties was said to have transformed the face of Indian Agriculture. The policies of funding the scientific community by the international foundations led to the evolution of High Yielding Varieties (HYV). The Indian Government was quick to recognise the potential of this technological development and give it full backing. However the spread of this variety was constrained by environmental factors. The benefit of the research was made available for the public welfare, and the public sector played a key role in the processes. Rights over varieties were not monopolised by any private entity nor were there any move to propertise the rights. The public interest was paramount. This is the reason why the Patent Law in India for a long time consistently excluded plant and plant varieties from the purview of patentability. But that is not the case with the advances in modern biotechnology. So the question which we in India need to ask and understand is why there has been this shift towards asserting property rights over inventions in the field of modern biotechnology as opposed to the public spiritedness that existed in the context of the green revolution.

More specifically in the context of this paper, the question can be framed as a problem of defining the factors that influence the shaping of the legislations in this field, and as a problem of identifying the players who have a role in the process of formulation of policy in this area. One reason for the increasing propertisation of the inventions of modern biotechnology is the increasing involvement of the private sector in research and development in this area. Private sector involvement can be traced in three phases after independence. The general trend appears to be a shift in spending with respect to R&D from public sector to private sector despite the fact that public sector industries are more numerous and engage in a greater number of research projects.

The Government of India, recognising the importance of interaction between private and public institutions, and hence in its Science and Technology Policy 2003, has declared as its objectives: "To encourage research and innovation in areas of relevance for the economy and society, particularly by promoting close and productive interaction between private and public institutions in science and technology". The Policy further declares that sectors such as agriculture...would be accorded highest priority and that key leverage technologies such as ...biotechnology.... would be given special importance.

## II Patent low regime

*Policy formulation*—The patent system is designed to encourage and maintain a continuous flow of inventions. The patent law governed the inventions in India since 1856. In India, the Controller of Patents, Designs, and Trademarks works under the Ministry of Commerce & Industry, Department of Industrial Policy & Promotion. The issue of patenting of agricultural biotechnology is a matter concerned with the Department of Industrial Policy and Promotion, Ministry of Agriculture, Department of Biotechnology, and the Ministry of Environment and Forests. The Department of Biotechnology may have to act as the nodal agency inviting opinion from various stakeholders in identifying the kind of protection the biotechnological inventions require. The Department of Industrial Policy and Promotion is concerned with the changes in the patent system.

The impact of the WTO TRIPS Agreement prompted the adoption of certain changes to the patent legislation. The Minister of Commerce and Industry, while presenting the Bill stated that Indian choice relating to TRIPS had been one of 'take-it-or-leave it', as part of a package of agreements. The obligation had to be balanced with many domestic interests. The decision of the Appellate Body of the WTO in the case between India and USA further had an impact on legislation in the country. Although the decision of the Appellate Body culminated in the first amendment to the Patent Act, several concerns were not answered. The issues that had to be addressed were brought forth in the second amendment. The amendment indicated the need for balancing international obligations with national interest. The minister identified the pillars on which the bill stands such as public interest, public health and nutrition, national interest, national security, protection of traditional knowledge and environment. He further said that the discovery of any living thing or non-living thing would not be patented. Similarly a micro-organism cannot be patented. However, the process by which the micro-organism has been developed can be patented if it meets the criteria for invention.

In debates over the bill to amend the Patent Act. the priority to be given to biodiversity protection was expressed by Shri Rupchand Pal. Mr. Mani Shankar Aiyar emphasised that there was need to ensure that genetic technology is not merely seen in terms of a profit motive but also that its social, developmental, and environmental consequences were fully taken into consideration. He referred to the example of *Bt* cotton and the long-term environmental loss that was likely to be caused apart from the immediate economic gains. Protection of bio-resources, and food security of the poor were key issues among the six critical points highlighted by Shri Subodh Mohite. The concern for protecting traditional knowledge and the issue of the influence of multi-national corporations (MNCs) were highlighted by Mr. Ajoy Chakraborty.

*Patenting of life forms*—Living things were not considered patentable in India in light of the public religious and moral sentiments. According to the multitude of Indian beliefs, such a right is vested only in God. Any departure would outrage the sense of morality of the Indian people. S.2(j) of the Patents Act 1970 provided that any 'substance produced by manufacture' was considered to be an invention provided it was new and useful. The Act expressly excluded from patenting "*a method of agriculture or horticulture; or any process for the medicinal, surgical,* curative, prophylactic or other treatment of animals or plants to render human beings or any process for a similar treatment of animals or plants to render *them*......".

The Act was silent with respect to patenting of living subject matter. Furthermore the Patents Act 1970 strengthened governmental control in what were viewed as sensitive areas in two important ways:

i. It rejected outright product patents in the field of drugs, food and outputs of chemical processes.

ii. It included provisions for compulsory licenses in the public/ national interest. Thus the 1970 Act balanced rights and obligations to ensure that patent monopolies were not established.

*Patent office directive*—In the absence of any express provision excluding from patentability living subject matter, there was no uniform interpretation of the provisions of the Patents Act by the different patent offices. As a consequence of this, the Controller General of Patents, Designs, and Trademarks in 1991 through an executive order took a specific stand that no patents should be granted in respect of living subject matter. Accordingly there was a bar on the patenting of the process or product of any life form including microorganisms, plants and animals, and parts thereof irrespective of how these inventions had been made. The instruction prohibited production of such substances by way of gene therapy, tissue culture, cell fusion, etc., from patentability. It allowed patents only for processes or methods of production of tangible and non-living substances such as enzymes, antibiotics, insulin; hormones, interferon, etc.

*The decision in Dimminaco case*—In the Dimminaco case the question of accepting a patent application for the process of creating a vaccine against Bursitis was answered in the negative by the Patent Office. The reason offered by the Patent Office was that section 2(j)(i) of the Patent Act precluded the word 'manufacture' from including the production of vaccines using micro-organisms and 'therefore' a process patent could not be granted to such a technique. Reference to the grant of patents for live cells, virus and micro-organisms by various branch offices was countered by the respondents on the ground that the micro-organisms in these applications had been lyophilized (freeze dried) and therefore could not be considered to be living. But this argument was not accepted by Court. The Court said that lyophilisation was in fact a method of preservation and 'therefore' indicated that the living organisms were in fact very much alive. The judgment has assumed importance since the Patents Amendment Act 2002 allows micro-organisms, but not other living substances, to be patented. Even after the Patent Act is given effect to it may be influenced by the interpretation in the Dimminaco case. There is no statutory bar to accepting a manner of manufacture as being a patentable invention even if it contains a living organism, provided the process is new and results in a useful product.

The judgment thus clarified that process patents could be granted though the end product contains living organisms and in spite of the fact that the process itself made use of micro-organisms as bioreactors.

## Obligations under TRIPS and India's Response

Biotechnology innovations are capable of satisfying the requirements for patentability. The crux of the patent debate, however, has never really been the ability of biotechnology to satisfy the requirements. The debate has centered primarily on whether we should reject patents for biotechnology for social reasons, regardless of their novelty, utility, and non-obviousness.

Article 27(2) of TRIPS is often cited as a means of justifying the exclusion from patentability of living organisms. Article 27(2) enables WTO members to exclude from patentability inventions, the prevention within their territory of the commercial exploitation of which is necessary to protect *ordre public* or morality including to protect human, animal or plant life or health or to avoid serious prejudice to the environment, provided such exclusion is not made merely because the exploitation is prohibited by their law. India has had a strong religious and cultural tradition which opposes patenting of life forms. It has therefore, been argued that Article 27.2 of TRIPS could be used to justify exclusion from patentability of living subject matter. The first attempt made by the government of India to amend the Patent Act in 1995, lapsed in Parliament. Consequently the US filed a case against India through the dispute settlement mechanism of the WTO. The WTO Appellate Body issued an adverse finding against India. In order to comply with the WTO ruling, the Government of India enacted in 1999 the Patents (Amendment) Act, establishing a mailbox facility to accept product patent applications from 1 January 1995 onwards, and to provide exclusive marketing rights (EMR) to such applicants. The amended legislation also provided for changes in the scope of patentable inventions, grant of new rights, extension of the term of protection, provision for reversal of burden of proof in case of process patent infringement, and conditions for compulsory licenses.

Section 3(j) of the Patents Act that had provided that no patents could be claimed for treatment of plants to render them free of disease or to increase their economic value has been amended as to exclude 'plants' from its purview. Thus patents can now be granted for a process for treatment of plants, which renders them free of disease or increases their economic value. The effect of the present amendment can be illustrated through the instance of a patent not being available for the cotton seed, which contains the *Bt* gene, while the process of engineering the gene into the seed would be patentable. This is because of the argument that the *Bt* gene is primarily a treatment, which makes the cotton plant more resistant to the bollworm (a pest to which it would otherwise be susceptible) and also increases its economic value. The process of achieving this result is patentable under section 3(i). Once a patent is granted, no other person would be allowed to utilise the same process to develop seeds with this gene, without a license from the patentee.

India has complied with its obligation under Article 27.3 (a) TRIPS by deleting the word 'plants' from section 3(i) which excludes 'any process for the medicinal, surgical, curative, prophylactic diagnostic, therapeutic or other treatment of human beings *or any process for a similar treatment of animals* to render them free of disease or to increase their economic value or that of their products' from the scope of patentability. India has also satisfied her obligation to give effect to Article 27.3(b) by incorporating Section 3(j), excluding plants and animals (in whole or any part thereof) other than micro-organisms but including seeds, varieties and species and essentially biological processes for production or propagation of plants and animals the exclusion in Section 3(j) from the scope of patentability. In addition it has given effect to the exception through the Protection of Plant Varieties and Farmers' Rights Act 2001 (see further below). On the whole, the obligation under Article 27.3 has been discharged by India. However, in the process, it diluted the idea of excluding completely plants from patent protection as appears to be underlying Article 27.3. This is because the net result of granting patent protection to treatment of plants is likely to end up in protection for the plant itself. It is questionable whether this dilution is favourable to biotechnological inventions or not.

In addition, section 5 of the Patents Act provides that no patent can be granted for any 'substance intended for use or capable of being used as food'. The rationale for this provision is that the first priority of the government is to ensure availability of adequate food to each and every Indian citizen. As this is one of the basic amenities in life, it would be against public interest to grant monopoly rights to individuals. However, the Act has narrowed the definition of 'food' by inserting the phrase 'for human consumption'. Under the earlier definition, 'food' was defined as 'any article of nourishment'. Thus, by implication the new definition excludes articles of nourishment not for human consumption, i.e. fodder for animals. Animal fodder is as important as food for humans in an agrarian economy like India, and patentability could cause great hardship.

Some of the NGOs have expressed the opinion that India should avoid patenting on micro-organisms by invoking the clauses of *order public* and offence to prevailing norms of morality. The Patent Amendment Act of 2002 is yet to come into force and is likely to come into force during March /April 2003.

## III Protection of biological resources

*Policy Formulation Process*—India has been playing an active role in the international conservation of biodiversity, as illustrated in its country report to the U.N. Commission on Sustainable Development. India's positive response to the principles enshrined in the Convention on Biological Diversity (CBD), relating to the rights over the genetic resources *per se* and to the technologies that are based on those genetic resources, is amply reflected in the Biodiversity Bill, 2002. This provides for the rights of traditional communities who have been the custodians of genetic resources, and have the knowledge to exploit them in a sustainable manner, measures to conserve and sustainably use biological resources, including habitat and species protection (such as declaration of biodiversity Heritage Sites), environmental impact assessments of all projects which could harm biodiversity, and integration of biodiversity into all sectoral plans, programs and policies. Regulation of access to biological resources by Indian nationals for conservation and to stop over-exploitation (e.g. of medicinal plants), while exempting local communities from unnecessary restrictions is also provided in the Indian bill. Even before the adoption of the CBD, the initiation of consultations on the protection of biodiversity came from the NGOs although the CBD does not explicitly require the participation of the non-governmental sector in policy making. The Ministry of Environment and Forests (MoEF) responded positively. The consultative process led to the establishment of an expert group that gave its report making recommendations for a Biodiversity Law in India.

The participation of NGOs was so significant that a Draft Law on Biodiversity was framed by Gene Campaign leading to effective National Consultation Seminars and the establishment of a committee under the leadership of MSSRF. The final version of the Biological Diversity Bill discussed and debated by the committee was released by the MoEF for public comments and wide ranging consultations were held with representatives of scientific and academic institutions, the Confederation of Indian Industry, the relevant government ministries and leading NGOs.

The Indian commitment was evident from its budget proposal of 1999-2000 for the proposed establishment of a National Bio-resources Board (NBB) as well as for the proposal to set up a National Innovation Foundation (NIF). The NBB consists of representatives from the Ministries of Environment and Forests, Agriculture, Department of Biotechnology and non-governmental experts. The object of mobilising intellectual property protection and converting innovations into viable business opportunities was sought to be realised through establishment of NIF.

Despite such concrete efforts, the Biodiversity Bill took a very long time to reach the take-off stage in view of political uncertainties at the central government level and lack of continuity of officials and functionaries in the concerned ministries. It is criticised that there was lack of political commitment to the formulation of law and action plan as well as divergent views between various participants made it difficult to arrive at a consensus.

*Emergence of biodiversity law*—The establishment of the Swaminathan Committee heralded a change in the focus of the Biodiversity Bill. The basic concern focused earlier on the rights of indigenous communities shifted towards the prevention of unauthorised use of biological resources by foreign individuals, institutions and companies. NGOs were apprehensive that the new focus would dilute the right to safeguard the knowledge of local and indigenous communities. The representatives from industry felt that finished products should not be included in the law and also suggested that the law should mandate Material Transfer Agreements. It is an indication of clear polarisation of interests between NGOs and industries. At the same time the Bill was considered to be a compromise between extreme views advocated by industries and NGOs. The proposed multi-stakeholder National Biodiversity Authority (NBA) has been conferred with the power of opposing the grant of patents on a substance process derived from the biological resources from within the country, The NBA is also provided with discretion to decide the extent of benefit sharing amongst the stake holders.

The Bill is problematic in its treatment of gene banks. Though the Bill provides for collaborative research it is understood to follow exclusionary policy under the Bill, and could adversely affect the development of gene banks such as the Consultative Group on International Agricultural Research (CGIAR). Such gene banks have an important role to play as they facilitate sharing of resources especially in the field of agriculture. These were the Centres which provided some of the green revolution varieties that had significant impacts on overall food production. Research undertaken in these centres is responsible for many scientific breakthroughs and the present policy of each country to preserve its biodiversity by asserting sovereign rights over its resources may have potentially disastrous effects, especially for the food security of the world. Cullet argues that a blanket ban on all foreign use may be counter productive.

Access to India's biological resources would enable small and less developed countries to foster the fulfillment of basic food and health needs. Further, such a provision will lead to adoption of similar provisions by other developing countries. However this argument is not tenable. Because, providing free access to these international research institutions may serve as a backdoor entry and there is the possibility of the multi-national corporations obtaining access in an indirect way. Further, the composition and nature of these institutions are also changing. The draft Biodiversity Bill reflects differential treatment for Indians and foreign persons seeking access. However, overall, the involvement of industry in the biodiversity law and policy processes was much limited.

The bill is not without shortcomings. Despite the fact that it was NGOs that took the initiative and they were the only active participants in the process of formulation of the policy and legislation, NGOs rarely stood united on various issues. There was lack of informed debate among the local and tribal communities as no consultations were held at the village level, being conducted instead in Delhi. Further there were different priorities that were emphasised by the different departments of the government and hence coordination in implementation would be the casualty.

In debates on the Bill, fears were expressed by the members in Rajya Sabha that the National Biodiversity Authority and the State Biodiversity Boards would be comprised mainly of bureaucrats and the defect of these two bodies was the overconcentration of bureaucrats, and whatever rights and authorities are given, were taken away by the central government. Saif-ud-din soz expressed fears that multinationals would be here to stay and they were concerned with commerce rather than India's future.

B.P.Apte strongly criticised the bill as not guaranteeing the protection of the community intellectual rights. He further expressed his concern that the bill itself would end up being a highway for those who would patent Indian traditional uses for two reasons—for not being vigilant and for functioning in the same bureaucratic manner.

On the whole, the consultative process adopted by the ministry in the formulation of the biodiversity law has been acclaimed as very progressive. It involved a high level of debate and generated a wealth of information. It is worth noting that there is no legal mandate in India, or any pre-determined mechanism, for consultation with all stakeholders affected by a law.

*Biosafety*—In 1989 the government of India issued the rules for the manufacture, use, import, export and storage of hazardous microorganisms, genetically engineered organisms or cells, which to date comprises India's biosafety law. These Rules needed updating pursuant to the Cartagena Protocol on Biosafety to the Convention on Biological Diversity, signed by India on 23 January 2001. There is an urgent need to bring the rules up to date with the international scientific knowledge, information and experience on biotechnology.

The government of India itself admitted in the second report to the CBD, there are not adequate mechanisms in the country to deal with potentially hazardous technology. For instance, open field trials of Monsanto's transgenic cotton have been allowed by the Department of Biotechnology without proper approval of the Genetic Engineering Approval Committee of the Ministry of Environment and Forests. Highlighting the possible risks to human and ecological health, as well as the need of clear jurisdiction in the biotechnology and regulatory system a writ petition was filed in the supreme court challenging these open field trials. The matter is still pending before the supreme Court. In the meantime, transgenic *Bt* cotton was found to be growing in the Western State of Gujarat late last year without the centre or the state governments having given permission for the same. With such an apparent bypass of the regulatory system, posing risks to the natural environment and divided centre and state opinions on the manner in which it should be dealt with, the debate on whether India should adopt transgenics in agriculture has been rekindled anew. There has been an aggressive propaganda by multinational agribusiness corporations selling genetically engineered crops/ products and by government circles in India. In the midst of this propaganda effort, several NGOs are continually stressing biosafety concerns.

The whole concept of the green revolution that revolved round germplasm conservation and its improvement through R & D, as well as the entry of private seed industry into the field, has functioned as the driving intensity in the formulation of seeds policy in India. The Patent Act did not specifically address seeds. The National Seeds Policy 2001 has endowed a framework for ensuring the growth of the seed sector in a liberalised economic environment. The object is to 'provide the Indian farmers with a wide range of superior seed varieties and planting materials in adequate quantities'.

The amended Patent Act provides that plants and animals in whole or in part, other than micro-organisms, but including seeds, varieties and species and essentially biological processes for production or propagation of plants and animals, are nonpatentable. Thus, even after the present amendment seeds are expressly excluded from the purview of patentable subject matter.

### IV Protection of plant varieties

*Background & Policy Formulation*—As noted above, plant genetic resources were not subject to any IPR protection in

India until recently. However, India was an active member of the FAO Commission on Plant Genetic Resources, which developed the International Undertaking on Plant Genetic Resources (IUPGR). This arrangement provided for the deposit of plant germplasm in international gene banks, which could be freely exchanged between countries. The main concern was that developed countries were exploiting the system for their advantage. Firstly, it was argued that most of the world's base crop collections and deposits of germplasm were held in the developed world, even though most of the accessions had come from the developing world. Secondly, while traditional varieties and farmers' varieties were treated as being the 'common heritage of humankind', the plant breeders in developed countries were securing IPR protection for their varieties (many a time developed from traditional varieties). In response to a decision to revise the international undertaking to bring it into line with the provisions of the CBD on access to genetic resources and benefit sharing, the Commission on Genetic Resource for Food and Agriculture adopted a revised text on 1st July 2001, which was submitted to the 31st session of the Food and Agriculture Conference. It was here that the International Treaty on Plant Genetic Resources for Food and Agriculture was adopted on 3rd November 2001. India signed and ratified the treaty on 10th June 2002.

Even though India did not provide IPRs over plant varieties, its own folk and traditional varieties, as part of the gene banks, were being used for research to develop new plant varieties in developed countries. The introduction of Plant Breeders' Rights (PBRs) into Indian agricultural dynamics should be seen in the above context. The Protection of Plant Varieties and Farmers' Rights Act, 2001 has been enacted in pursuance to Article 27.3(b) of the TRIPS Agreement which requires WTO members to provide for the protection of plant varieties either through patents or through a *sui generis* regime or a combination thereof. The Patents Act 1970 excluded plant varieties from the purview of protection. Thus TRIPS required India to put in place a *sui generis* regime for protection for protection of plant varieties. The initial draft Plant Varieties Bill, which was introduced in Parliament in December 1999, had generated a lot of criticism among NGOs and farmers' lobbies. A Joint Parliamentary Committee was then appointed in 1999. This Committee visited 15 states and recorded oral evidence of representatives of farmers, experts, individuals etc. and received 132 memoranda containing suggestions. However, certain quarters have criticised this process as being non-transparent and conducted in a very hurried manner.

The then agriculture minister, while introducing the Protection of Plant Varieties and Farmers' Rights Bill in the Parliament, explained thus: The concept of plant breeders' rights arises from the need to provide incentives to plant breeders engaged in the creative work of research which sustains agricultural progress through returns on investments made in research and to persuade the researcher to share the benefits of his creativity with society. A system of plant breeders' rights encourages better and mission-oriented research for development of varieties that are fully suited to a given agro-climatic region.

India has developed commendable strength in agricultural research. Indian breeders working, mainly, in the public research system have developed a large number of new varieties. In the absence of plant breeders' rights, these varieties would be freely available to others for exploitation. New varieties developed on the basis of these varieties could get protected in other countries without any benefit accruing to Indian institutions/organisations, whereas the availability of varieties developed in countries which provide for plant breeders' rights would be restricted in India. Therefore, putting in place a system of plant breeders' rights through law in India provides protection to the plant varieties developed by public research system. A system of the plant breeders' rights in the country would also encourage foreign companies to organise buy-back production of seeds in India for export to their countries without any fear of unauthorised use of their genetic material.

The Indian parliament passed the Protection of Plant Varieties and Farmers' Rights Act (PPVFRA) in 2001, but, at the time of writing, it is yet to come into force. The debates in the Upper House of the Indian parliament over the bill indicate a deeprooted concern about the hasty manner in which the bill was being passed, especially in the context of India still having time, until 2005 under TRIPS, for setting in place a regime for protection of plant varieties. The debates also indicate that there was considerable concern expressed by members belonging to a wide political spectrum on the detrimental effect that a law in its present form would have on the farmers and Indian agriculture.

Among the two schools of thought in India regarding adoption of *suĭ generis* plant variety protection, one school favours the adoption of a UPOV model of plant varieties protection, whereas the second school advocates a non-UPOV frame work for protection of Breeder's Rights which could also uphold rights of local communities conserving the germplasm which forms the foundation of protection of plant varieties. The second school is associated with green movement in India. India being a rich mega diversity country and having a rich storehouse of land races of principal agricultural crops and also because it has strong R&D base in conventional methods of plant breeding methods, has adopted a *sui generis* protection system in the PPVFRA.

Agricultural research in India has been characterised by public funded research. The green revolution in the 1970s was spurred by public sector research and thus the high yielding varieties, which were developed, were available to all farmers. With the introduction of PBRs, there would be exclusivity in agricultural technology and poor farmers might be bypassed by the technological changes.

*Protection of plant varieties and farmer's rights act—* The Protection of Plant Varieties and Farmers' Rights Act has substantial aspects derived from the International Union for the Protection of New Varieties of Plants (UPOV) at the same time departing from it in certain fundamental ways as will be described below. This was possible because India is not a signatory to the UPOV and the TRIPS agreement too did not mandate for accession to the UPOV. Thus, India was able to use this flexibility and incorporate the many safeguard provisions and positive farmers' rights into the Plant Varieties Act.

However, the Indian government recently announced its intention to sign the UPOV. The official reason put forward by the government for signing the UPOV is that doing so would guarantee international recognition for Indian plant varieties and that it is also necessary to acquire skills and material resources for plant breeding. The minister, while replying to a question stated on the floor of Rajya Sabha in July 2002 that joining UPOV would be in the interest of our farmers as this would, *inter alia*, facilitate greater investment in research for the development of new plant varieties; in order to ensure the availability of quality seeds to farmers; obviate the need to enter into a large number of bilateral agreements with other countries for mutual recognition of plant breeders' rights; enable Indian plant breeders to secure protection in all conventional countries with minimal formalities and costs, etc.

If India does indeed become a member of the UPOV, there would be substantial repercussions. This is because although the PPVFRA, 2001 follows the same model of protection as that of the UPOV, many safeguards and balances have been incorporated into the Act, such as provisions relating to farmers' rights that might run contrary to UPOV.

It is criticised that this move appears to be in response to international pressure and pressure from the powerful commercial plant breeders. Activists have criticised the view of the government. Suman Sahai of Gene Campaign commented: "The argument presented by the government is not only bogus, it is shameful- that joining UPOV was necessary to acquire skills and material resources for plant breeding. If the scientists of the ICAR are so stupid that we need outside skills, let us first close down ICAR. If the intention is to give unlimited rights to biotechnology companies like Monsanto at the cost of Indian farmers, let the government make a statement in parliament that that is the purpose of joining UPOV. In all the arguments presented by the government, the word 'farmer' finds no mention. It is only the 'breeder' and the 'company' whose rights and interests are discussed".

*Plant breeders' rights*—The criteria for protection and the rights given to plant breeders under the PPVFRA and the UPOV 1991 are very similar. To enjoy protection, the variety should conform to the criteria of being novel, distinctive, uniform and stable and the rights, which accrue on protection, include the exclusive right to produce, sell, market, distribute, import or export the variety. The scope of rights under the UPOV 1991 are wider to the extent that the breeders' rights extend over the harvested material and the products made from the harvested material which are obtained from the unauthorised use of the propagating material. Indian law does not have any such provision. India prefers the less stringent version of UPOV 1978 whereby Plant Breeders' Rights are conferred only over 'reproductive' and vegetative propagating materials of the protected variety.

Varieties capable of enjoying protection under the PPVFRA include extant varieties (i.e., varieties pre-existing the commencement of the Act), farmers' varieties, essentially derived varieties and new varieties. These rights accrue upon registration. It is noteworthy that under the UPOV extant varieties and farmers' varieties do not enjoy protection. The importance of recognising extant varieties as being capable of protection lies in the fact that India has a rich variety of plants being traditionally produced, and which pre-exist the coming into force of the act. It would 'therefore' afford a mechanism of protecting these varieties. Even if India does become a party to the UPOV, there would be no international recognition of protected extant varieties, registered in India. Hence, the provision of protecting extant varieties would not be contrary to the UPOV The maximum duration of protection under the Plant Varieties Act currently is 18 years in the case of trees and vines, 15 years from the date of notification in the case of extant variety and 15 years from the date of registration in the case of a new variety. This is in sharp contrast to the UPOV 1991, which requires much longer terms of protection. The concern has been that after a reasonable duration of protection, the variety should be available to be freely used by farmers. To this extent, India preferred to adopt Article 8 of UPOV 1978.

The scope of Plant Breeders' Rights (PBRs) under the Act is to some extent diluted (compared to the UPOV 1991) also by the nature of farmers' exemption and the concept of farmers' rights as is embodied under the Act. Thus, we see that an effort has been made to balance the interests of farmers, and commercial plant breeders while delineating the scope of plant breeders' rights under the Act.

*Farmers' rights*— International recognition of farmers' rights is found in the International Treaty on Plant Genetic Resources for Food and Agriculture. The PPVFRA confers on farmers the right to register a variety that has been developed or bred by the farmer in a like manner as a breeder of a variety. The rationale for this is to prove the point that monopoly IPRs are not restricted to large seed companies and commercial plant breeders, but can be acquired by ordinary farmers as well. It is difficult for farmers' varieties to fulfill the criteria of being stable and uniform. Thus, while technically farmers' varieties can be protected and farmers can acquire IPRs over their varieties, the system of PBRs which has been developed to protect varieties developed in laboratories, is inherently unsuitable to protect varieties which develop and evolve *in situ.* UPOV does not contain a similar provision.

Farmers who are engaged in the conservation of genetic resources of land races and wild relatives are entitled for reward for *in situ* conservation and preservation. These farmers would be entitled to benefit sharing under Section 26 of the PPVFRA if such varieties have contributed to the development of a new variety. However, this provision has been criticised on the ground that it would be very difficult to secure such benefits as the process of placing a claim and establishing entitlement is a very burdensome. A majority of Indian farmers may not be in a position to do all this given their economic and educational status.

Farmers are also entitled to compensation from breeders if the propagating material sold to the farmers does not show the expected performance in given conditions which have been disclosed by the breeders. However, the importance of such a provision is underlined by lessons from past incidents. During the late nineties, farmers in Andhra Pradesh, Karnataka and Maharashtra borrowed huge amounts of money for using seeds and pesticides in the cultivation for export purposes. However, bad quality seeds led to huge crop failures. What followed was large-scale indebtedness and suicides due to non-repayment of loans. The vulnerability of the small Indian farmer, especially to the risks of producing cash crops for the export market is clear. This seems to be the motivation behind this provision. However, it has considerably irked the industry as being too onerous. The UPOV 1991 has no such provisions and in the event that it joins UPOV, India might be asked to revise this since they could be interpreted to be violative of the breeders' rights.

*Farmers' exemption & innocent infringement*—A farmer is entitled to save, use, sow and resow, exchange, share or sell his farm produce including seeds which would be protected under the Act, provided that the seed is not branded. This provision thus recognises and legitimises the practice of trade in seeds (even protected seeds) between farmers in villages and towns, prevalent in India for centuries, and to some extent dilutes the monopolistic stronghold of the PBRs by diluting the exclusive right to sell the protected seeds. Exchange of seeds between farmers allows them to plant and grow crops they want to cultivate by trading one variety of seed for another. This provision is especially beneficial for poor farmers, as it enables them to borrow or buy seeds from other farmers cheaply. Moreover, exclusive marketing rights for the breeders or seed companies would have promoted crops, which are commercially most viable, gradually effacing many traditional varieties, and methods of mixed cropping. Thus, this provision is of crucial importance even though it is merely declaratory. Farmers' exemption under the UPOV does not provide for the sale of the propagating material.

Section 42 of the Act protects innocent infringement, i.e., when the farmer was unaware of the existence of the right. Though the rule under the PPVFRA is open to subjective interpretations and arbitrary rulings. It also places the burden on the farmer to prove that at the time of the infringement, he was unaware of the existence of the right. Despite these shortcomings, it still remains a genuine attempt to understand the situation of the farmers and their vulnerability.

*Rights of communities*—In addition to farmers' rights, rights of communities have also been given some recognition. Any claim attributable to the contribution of the people of that village or local community, in the evolution of any variety for the purpose of staking a claim on behalf of such village or local community, can be made to the authority, which on finding such a contribution to be 'significant', will notify a certain amount to be paid as compensation. However, the burden of claiming benefit sharing or compensation is on the claimant. Concerns have been expressed about whether the information about the registration of a given variety will reach the communities or farmers concerned, especially since claims can only be filed at the post registration stage. This is a very important provision for a gene-rich country like India. In India, despite a lack of clarity and cumbersome nature of the benefit claiming process, this rule clearly recognises community rights and attempts to prevent biopiracy.

*Public interest provisions*—The UPOV convention does not contain any provision relating to compulsory licensing or exclusion of varieties. However, Article 17 of the UPOV lays down that no restriction other than those expressly laid down in the UPOV Convention may be imposed on the rights of the breeder other than in the name of public interest. Thus, these provisions might be sustained on this ground even if India does become a party to the UPOV.

1. Compulsory licensing: The UPOV does not contain any provision for compulsory licensing. The grounds based on which a compulsory license can be granted are:

- When the *'reasonable requirements' of the public* for seed or other propagating material have not been satisfied; or
- The seed or other propagating material of the variety is *not available to the* public at a reasonable price.

In the above circumstances, a compulsory license can be granted for the production, distribution and sale of the seed or other propagating material of the variety. As shown in the Table, the UPOV does not contain a similar provision. This provision is particularly relevant in a country like India in the context of drought conditions.

2. Exclusion of certain varieties from protection: It is provided that no registration shall be made under the Act in cases where prevention of commercial exploitation of such variety is necessary to protect public order or public morality or human, animal and plant life and health or to avoid serious prejudice to the environment.

This provision is extremely important in view of the fact that a number of crop varieties of rice, wheat, pulses and other food crops are essential to protect the food security of the country. It has been suggested that this provision must be effectively used in order to ensure that the varieties, which constitute and cater to the food needs of the poor, are not afforded protection. Moreover, under the Indian law, registration of a plant variety is not allowed if the variety in question involves any technology such as Genetic Use Restriction Technology (Terminator Technology) which is injurious to the life or health of humans, animals or plants. India, in adopting PPVFRA has chosen the path of *sui generis* plant varieties protection. It is viewed that UPOV 1978 is more suitable to Indian situation in terms of 'Breeder's exemption' and 'farmers' privilege' than UPOV 1991, and also because Indian breeders have developed new cereals and non-cereal varieties. Even if UPOV allows accession to 1978 version, developing countries would be pulled into UPOV process. Reading Article 31 and Article 14 of UPOV 1991, it is possible for a breeder from a UPOV 1991 country to authorise export of his protected plant varieties to a researcher in a non-UPOV country. The researcher can generate an essentially derived variety. UPOV 1978 and UPOV 1991, not being compartmentalised regimes, UPOV 1991 is bound to predominate over its preceding 1978 and 1961 versions.

*Views of stakeholders*—The NGOs and activists have expressed the view that India is emulating the biotechnology policy of UK, US, EU in formulating her policy and law on biotechnology. India also seems to be influenced by international institutions like the UN agencies/donor agencies who have used the carrot of join collaboration in research and also foreign direct investment.

As regards the question of *adequacy* of the protection in respect of India's economic and social interest, it was opined that these policies and law are not adequate to shield India's farmers and biodiversity. While India's IPR regime is in tune with the international requirements, what is not acceptable is that such an IPR regime will ultimately negatively impact the country's economy. NGOs demand a framework that protects India's economic and social interests. Since the biotech industries are profit driven and they focus on producing commercially lucrative transgenic crops, and not necessarily those crops which may be less lucrative but aim at improving the nutritional status of the citizenry. Referring to the case of cassava, an activist stated that despite it being a staple food for at least 300 million people in Africa, no biotechnology company made any effort to improve the crop yield and production.

One activist further stated, "Moreover, the tall claims by the seed companies are not always true. The United States Department of Agriculture and the US government having been pushing the transgenic crops into the third world claming that they have been tested for safety in their country. However, this is not a guarantee against technological failure or environmental disaster. It is a well-established fact that the extent of hunger and malnutrition that prevails in India is not due to lack of food production but is the result of yawning gaps in reaching food to the vulnerable sections of the society.

Responding to the question of how they evaluate the *role of current IPR* regime, a number of respondents said that it only reinforces the control of the multinational industries over the seed. IPR protection will increase the price of seeds and this would put them beyond the reach of poor farmers. Further the consequent increase in seed price will also put the food grains beyond the purchase power of the ordinary consumers. In fact, India has no mechanism to ensure that the national interest for biotech research remain safe under the IPR regime that is perceived as having been thrust on India. In response to a question about the UPOV and the present Protection of Plant Varieties and Farmers' Rights Act 2001, several NGOs and activists expressed their opinion, that there are two categories of IPRs that have a direct impact on the erosion of prior rights of communities: patents and plant breeders' rights. Plant breeders' rights negate the contribution of third world farmers as breeders and hence undermine farmers' rights. Patents allow the usurpation of indigenous knowledge as a western invention through minor tinkering. The UPOV Convention is seen as a Western device by Indian NGOs, which along with patenting leads to biopiracy. This form of intellectual property rights protection, referred to as a plant breeders' rights, is being promoted as the most favourable form of adoption under the *sui generis* option for developing nations by the developed nations. But according to the 1991 revision of the UPOV Convention, newly introduced clauses severely restrict farmers' rights by removing all rights for them to save seed for sowing for the following year, as well as removing researchers' rights to save the seed of new protected varieties. The protected variety may still be used as an initial source of variation for the creation of new varieties but such new varieties cannot be marketed or sold without the plant breeders' rights' holder allowing it. Further, UPOV is a monopoly system that embodies the philosophy of the industrialised north who want to protect the interests of corporate biotechnology and powerful seed companies. If India does not evolve its own *sui generis* system centered on community intellectual rights of farmers and adopts the UPOV model, a rights regime will have been created that protects the rights of the seed industry but offers no protection to the rights of farmers. This in turn will allow a free flow of agricultural biodiversity based on centuries of breeding from the fields of Indian farmers, while the farmers have to pay royalties to the seed industry for the varieties derived from farmers' varieties.

It was further opined that the UPOV system is not in India's interest for the following reasons:

Firstly, it is too expensive. The cost of testing, approval and acquiring a UPOV authorised Breeder's Rights Certificate will cost about two to three lakhs at least. This could even go upto ten lakh rupees. Such high rates preclude the participation of all but the largest seed companies. It will be unaffordable for any small company, farmers' co-operative or farmers or breeders. Secondly, UPOV cannot be accepted by a developing country like India because it is based on the philosophy of the industrialised nations where it was developed and where the primary goal is to protect the interest of the powerful seed companies who are the breeders. In UPOV, rights are granted only to the breeder, and there is nothing for the farmer.

Thirdly, UPOV laws are formulated by nations, which are industrialised, not agricultural economies. In these countries the farming community is by and large rich and constitutes about 2-7 percent of the population. These countries do not have large number of small and marginal farmers like India. In Europe, agriculture is purely a commercial activity, but in India it is a source of livelihood. These farmers are the ones who have nurtured the genetic resources, which the breeder wants to capture under breeder's rights.

On the question of free trade and liberalisation, the response was that IPR regimes in the context of 'free trade' and 'trade liberalisation' become instruments of piracy in three ways: Firstly, by *resource piracy* in which the biological and natural resources of communities and the country are freely taken, without recognition or permission, and are used to build up global economies. For example, the transfer of *basmati* varieties of rice from India to build up the rice economy of the US; the free flow of neem seeds from the farms, fields and commons to corporations like W. R. Grace for export. Secondly, by *intellectual and cultural piracy* in which the cultural and intellectual heritage of communities and the country is freely taken without recognition or permission and is used for claiming IPRs such as patents, and trademarks even though the primary innovation and creativity has not taken place through corporate investment. For instance, the use by US corporations of the trade name 'basmati' for their aromatic rice. Thirdly, through *economic piracy* in which the domestic and international markets are usurped through the use of trade names and IPRs, thereby destroying local economies and national economies where the original innovation took place and hence wiping out the livelihoods and economic survival of millions. For example, US rice traders usurping European markets; race usurping the US market from small-scale Indian producers of neem based biopesticides. In response to a question on bio-resource management policy in India today an activist said that at present India's biotechnology policy seems more focused on safeguarding the interests of the various stakeholders and protecting India's rich resources. Thus, the Biodiversity Bill, Traditional Knowledge Bill, etc are primarily concerned with the conservation of the knowledge and resources indigenous to India and seek to ensure that traditional users of such knowledge are made 'benefit sharers' in any commercial exploitation of such knowledge.

It is obvious that the policy formulation process in the Indian intellectual property regime for agricultural biotechnology is characterised by a lack of cohesion among policy objectives and the absence of a coordinated effort to facilitate the balancing of the interests involved in the process, which are varied and often conflicting. The lack of coordination is characterised by the very effort to introduce legislative instruments premised on policy goals and conceptual foundations that are overlapping. The Ministry of Agriculture has drafted the Protection of Plant Varieties and Farmers' Rights Bill; the Ministry of Environment, the Biodiversity Bill; the Ministry of Science and Technology, the

Patents (Second Amendment) Act, and the Ministry of Human Resource Development has been attempting at formulation of the Traditional Knowledge Protection Bill. Each of these instruments evidences a certain dominant interest, which is sought to be protected under the regime in question. The Protection of Plant Varieties Act emphasises the creation of property rights entitled the plant breeder's right aimed primarily at protecting commercial interests such as those of large seed companies and commercial cultivators. The regime does also seek to provide for a minimal form of protection for farmers engaged in traditional forms of innovation. Innovations such as these would also form an integral component of India's biological diversity, a subject that is sought to be protected under the Biodiversity Bill. In addition to these, a completely independent regulatory regime for traditional knowledge is being mooted (though there does exist some confusion regarding the exact shape such a regime should take).

In situations where co-ordination among government agencies is found to be insufficient, it is unrealistic to expect a high degree of co-ordination between government agencies and private agencies and private entities. Most private stakeholders are of the opinion that there are some ministries in which a specific interest is given precedence over others.

The range and diversity of interests in the Indian biotechnology sector is one of the reasons for this shortfall. The dominant players include members of the industrial and scientific communities, NGOs and, by far quantitatively the largest, the farming communities. Not all of these interests are adequately represented in every policy formulation process. Nevertheless, each does find representation in the overall system. The net effect is a profusion of policies within the legal system, many of which are incapable of mutual co-existence.

At the outset, while it may be difficult to balance some of these interests, it nevertheless is essential to at least attempt to do so by bringing them together. Allowing each interest to function independently would be detrimental to India's national development and more importantly, to its bargaining strategy in international organisations.

While it may be beyond the scope of the present undertaking to suggest a concrete process by which this can occur, it may not be out of place to make a few generalised observations on the shape that an ideal, coordinated and transparent policy formulation process should take.

- The formulation process should be spearheaded by an Inter-Ministerial Coordination Committee on Intellectual Property consisting of representatives of every central ministry.
- The formulation process on any specific IPR related issues shall be initiated by the concerned department / ministry but left under the control of Inter- Ministerial coordination Committee on Intellectual Property. This would ensure that no interest is promoted over others and that free debate occurs in the process.

The effect of such process would be:

a) The creation of cohesive policy frameworks for different sectors and a common policy framework around a legal area, i.e. intellectual property.

b) Enhanced co-ordination neutrality and transparency in the formulation process.

c) An independent process for new interests and

stakeholders to participate without having to lobby for political support.

The patent law has still given priority to the community interests rather than to the monopoly private rights. By excluding from patentable inventions the life forms and inventions based on traditional knowledge, considering traditional knowledge or non-reference to the country of origin as grounds of opposition, Indian patent law has favoured keeping certain interests in the public domain. The idea of a compromise between the extremes of public domain and private monopoly seems to be the option that is being favoured here.

The concern of the State in safeguarding the genetic resources through establishment of national authorities and benefit sharing processes is on the anvil. Though the role of the proposed authority is confusing, controversial and overlapping, and at the same time the extra territorial jurisdiction of the authority is doubtful, a committed step in this direction is a positive sign.

Farmers' rights and rights of communities are regarded as a counterbalance to the introduction of private, monopoly IPRs into Indian agricultural dynamics. However, many concerns have been expressed as to whether in reality these provisions will actually benefit farmers and communities and tilt the balance in their favour. This is especially so in the context of India where majority of the farming community is poor and illiterate.

# Index